“HISTORY IS THE SOUL OF MAN”

Nicholas Deely

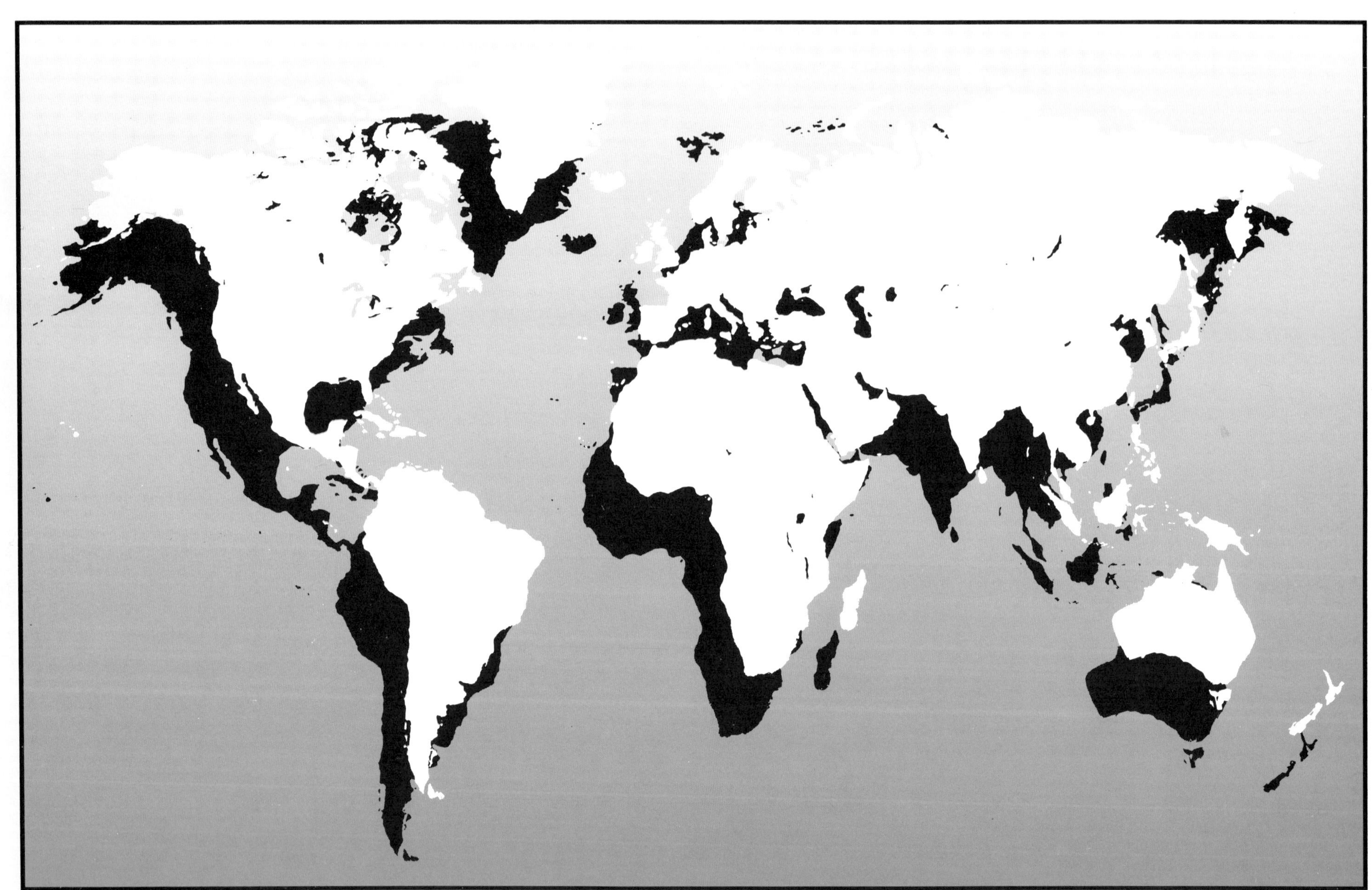

TANANA VALLEY RAILROAD
THE GOLD DUST LINE
ALASKA

Nicholas Deely

by Nicholas Deely

Dedication / Acknowledgements

Man is not an isolated entity, but rather one who needs to work in harmony with others to be creative and to be able to transfer dreams into a reality. While the concept of this treatise was developed from my own personal interests, it was only with help, co-operation and encouragement of so many fine people that the story of the Tanana Valley Railroad could have been chronicled and documented for posterity. To the people listed below, I dedicate this book:

"Bud" Anderson
Joe Ashley
Ollie Backlund
Richard Barlow
Jim Blasingame
Jim Brown
Terrence Cole
Bernice Deely
Dwight Deely
Pat Durrand
Dick Farris
Jeffrey Fay
Shirley Franklin
Curt Fontenberry
Paulo Greer
Dan Gullickson
Bruce Halderman
Steve Harmon
John Henderson
Jim Hutchinson
Nilo Koponen
Leslie Labrenz
"Clutch" Lounsbury
George Lounsbury
Joe Lounsbury
Colin MacDonald
M. Maddox
Evolyn Melville
Carl Mulvihill
Marge Naylor
Sue Dohse
Roy Olson
Robin Russell
Frank Ryan
George Simpanen
Olga Steger
Dory Stucky
Paul Solka
Richard "Tark" Tarkiainen
Robert Turnbull
Richard Veazey
Joe Vogler
Candy Waugaman
Dorothy Wilde
George Yurkovich

Archives, University of Alaska Fairbanks, Alaska.
Archives, Noel Wien Library Fairbanks, Alaska.
Anchorage Museum of History and Art Anchorage, Alaska.

CHAPTER 1

GENESIS

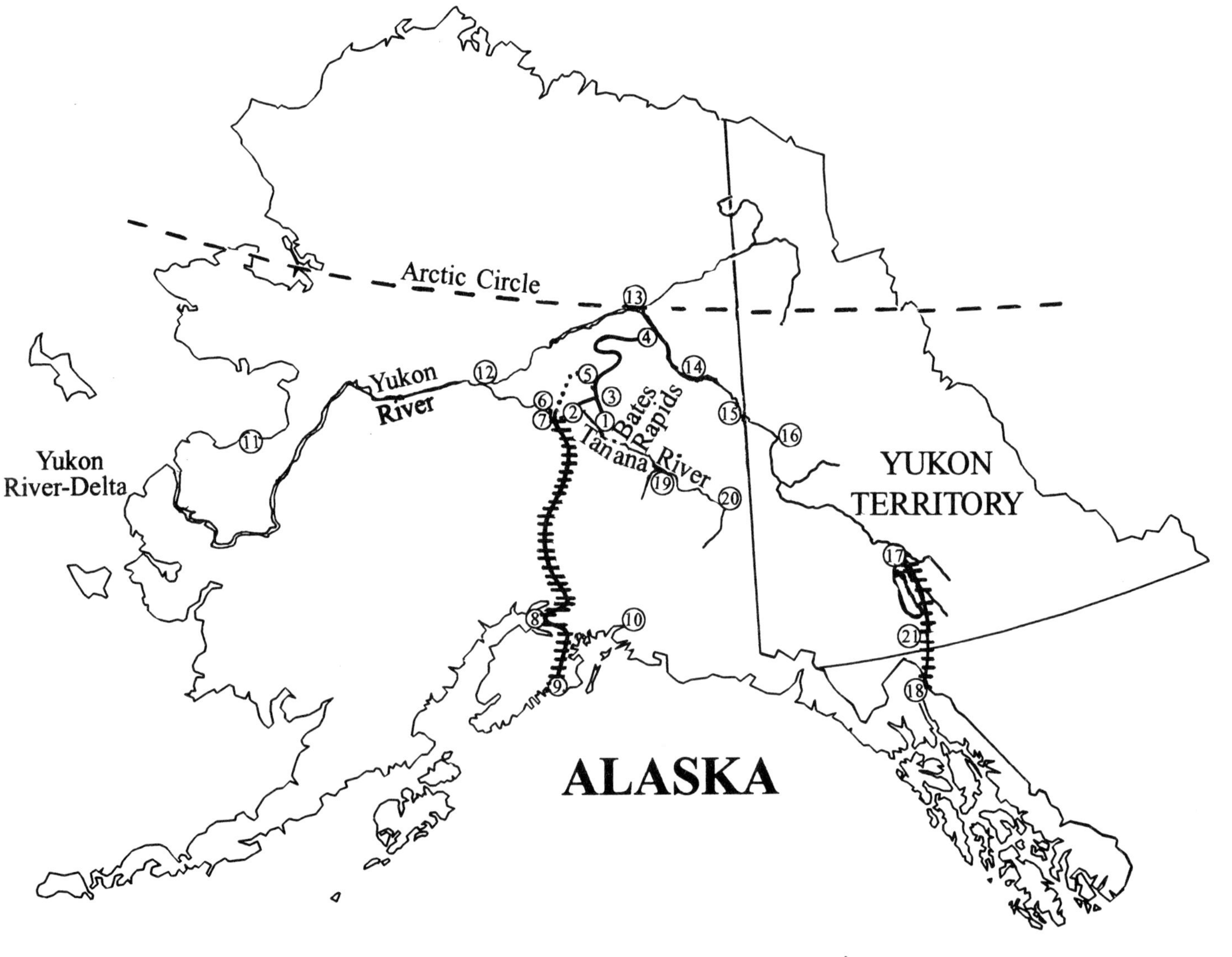

1 Fairbanks
2 Chena
3 Chena Junction
4 Chatanika
5 Sheep Station (Happy)
6 North Nenana
7 Nenana
8 Anchorage
9 Seward
10 Valdez
11 St. Michael
12 Ft. Gibbons (Tanana Village)
13 Ft. Yukon
14 Circle
15 Eagle
16 Dawson (Klondike)
17 Whitehorse
18 Skagway
19 Big Delta
20 Tanancross
21 White Pass and Yukon Railroad

Alaska Railroad — AEC Railroad — Tanana Valley Railroad

In the beginning, ALASKA was the land of *"The People"* — the Indians, the Inupiats (Eskimos), and the Aleuts. It was called ALYESKA — "The Great Land". Prior to 1741, the advent of the arrival of Vitus Bering, a Dane who was in the service of the Czar of Russia, it was a wilderness known only to *"The People"*. Consistent with the colonization ethic of the European nations at that time, it was summarily annexed by Russia, and renamed Russian-America. And — with this subjugation came a transition which has forever changed the face of Alaska.

In time, the city of Fairbanks was founded. Located 120 miles south of the Arctic Circle, at 64° 50' 45" north latitude and 147° 43' 15" west longitude, it is directly in line with Hawaii. At 434 feet above sea level, Fairbanks is both the beginning and the end of cosmopolitan North America for it sits alone in the vast wilderness of the sub-arctic. Resting on the floor of the Tanana Valley, in the interior of Alaska, it breaks dramatically from the archetypical picture of the frozen north. Fairbanks is not located in a setting of glaciated, sawtooth mountains, but rather in a region of lakes and rivers which speckle the lush green hills that surround it on three sides. To its south flows the mighty, silted Tanana River, full of glacial debris. This body of water, in conjunction with the adjoining snow-fed Chena River, forms the focal point upon which this saga of northern rails is built.

The surrounding hills, reaching heights of approximately 2,000 feet, are composed of schists and gneisses of highly metamorphosed rocks derived from sediments and lavas. Along many of the south-facing slopes can be seen mounds and layers of *Loess*. This material is composed of dust which developed from the grinding action of glaciers on the underlying rock. Through the years, this material was blown northward across the Tanana Flats from the Alaska Range and established distinctive new contours to the surrounding hills.

The topsoil which covers this entire region supports a heavy vegetative growth. Except in the discrete patches of stunted Black Spruce and Sphagnum Moss, which generally signify areas of discontinuous permafrost, this region has a prolific growth of both coniferous and deciduous trees, giving it an appearance much like that of northern New England. White Spruce, Black Spruce, Tamarack (Larch), Birch and Aspen, and a multitude of other trees and berry bearing shrubs drape the shoulders of these hills.

While its rivers and lakes sparkled in the sun, the earth below glittered with ***Gold***, more gold than has ever been seen in one place in North America! And with this gold came a form of madness which drives mans body and soul into a state of absurdity and extreme behavior not equaled by any other drive except those of survival and sex! Therefore, to fully understand the origins of the Tanana Valley Railroad, one must understand the genesis of the land, *"The People"* who were already here, and those who came later.

Author's Note of References

Excerpts from the following list of publications are used throughout this book. These references are recommended reading for any interested in Alaskan history.

E. T. Barnette, by Terrence Cole, Alaska Northwest Publishing Company, Anchorage, Alaska, 1981. ISBN # 0 - 88240 -154 - 8

Mining Railways of the Klondike, by Eric L. Johnson, Pacific Coast Division, Inc., Canadian Railroad Historical Association, Vancouver, B.C., 1995. ISBN # 0 - 9697633 - 4 - 4

Rails North, by Howard Clifford, Superior Publishing Company, Seattle, Washington, 1981. ISBN # 0 - 87564 - 536 - 4

Landscapes of Alaska, edited by Howell Williams, Berkeley and Los Angeles, University of California Press, 1958. L.C. # 58 - 8655

Alaska Science Nuggets, by Neil Davis, published by the Geophysical Institute, University of Alaska, Fairbanks, Alaska, 1982. ISBN # 0 - 915360 - 02 - 0

Steamboats on the Chena, by Hendrick and Savage, Epicenter Press, Inc., Fairbanks, Alaska, 1988. ISBN # 0 - 945397 - 00 - 3

Railroad In The Clouds, by William H. Wilson, Pruett Publishing Company, Boulder, Colorado, 1977. ISBN # 0 - 87108 - 510 - 0

Flag Over The North, by L.T. Kitchener, Superior Publishing Company, Seattle, Washington, 1954. L.C. # 54 - 12521

Along Alaska Trails, by Lois McGarvey, Vantage Press, New York, New York, 1960.

Fairbanks, $200 Million Gold Rush Town, by JoAnn Wold, Fairbanks, Alaska, 1971.

Bibliography continued on next page.

Paragraph 1 references *Steamboats On The Chena*, pg. xi
Paragraph 3 references *Landscapes of Alaska*, pg. 77

The Development of Settlement In The Fairbanks Area, Alaska — A Study Of Permanence, by Robert Leonard Monahan, Thesis — Graduate Study, McGill University, Montreal, Canada, April, 1959.

The Way It Was, by JoAnn Wold, Alaska Northwest Publishing Company, Anchorage, Alaska, 1988. ISBN # 0 - 88240 - 316 - 8

This Old House, by JoAnn Wold, Alaska Northwest Publishing Company, Anchorage, Alaska, 1976. ISBN # 0 - 88240 - 069X

Alaska Railroad, by Bernadine Prince, Ken Wray's Printing Shop, Anchorage, Alaska, 1964.

Old Yukon, by James Wickersham, Washington Law Book Company, Washington, D.C., 1938.

Pacific Northwest Quarterly, Volume 1, 1906 - Volume 53, 1962. Originally issued as The Washington Historical Quarterly, the name changed to Pacific Northwest Quarterly, Volume 27, January, 1936. Published by University of Washington, Seattle. L.C. # 8 - 30966

Alaska-Yukon Magazine, The Harrison Publishing Company, Seattle, Washington, September 1906 to July 1912.

Much history is preserved in newspapers and many of the anecdotes in this book were gleaned from early newspaper articles. However, it should be noted, that the standards we expect today — such as accuracy and truth in reporting, were not always upheld by earlier newspapers. In Alaska's pioneer days, *anything* was "news" and stories were often exaggerated or even fabricated to add "color" to the news. Early Alaskan mining camp newspapers served a highly transient population. They had no subscription lists, and made no attempt to obtain any such. Competition was fierce, and in 1907, Fairbanks reached an all-time high of eleven newspapers.
Typos were numerous and the articles reproduced for this publication contain the original misspellings.

Paragraph 4 references *Steamboats On The Chena*, pg. xii.

Little was known about the land and the people of the upper Yukon and Tanana Rivers, even into the latter part of the nineteenth century, except that it was a mysterious land inhabited by "unfriendly Indians". What little was known of these people was a result of the isolated and fragmentary communications between the Indians and the Russians. Contact with the people of the Interior was not of the magnitude as that experienced by the Native-Americans living along the coast of Alaska. Although a trading post had been established in the Interior by the Russians, their tenure was to end on October 18, 1867, when Russian-America was transferred to the United States for the sum of $7,2000,000. This payment was to be more than matched by the value of the gold that was extracted from the Fairbanks creeks in just one year — *the year of 1909*!

While the Russians were exploring the lower Yukon River, there began a westward migration of explorers from Canada which penetrated the region of the upper Yukon River. These bilateral excursions ultimately lead to the establishment of fur trading posts by the Russian-American Company, the Hudson's Bay Company of Great Britain, and the Western Fur and Trading Company. As with the French-Canadian *voyageurs* river treks across Quebec, Ontario and Minnesota, these explorations along the upper Yukon were carried out by canoe and other small river craft.

The Tanana River was first explored by the white man sometime between the years of 1874 and 1878. Early explorers crossed overland from a point near Eagle on the Yukon River, which is located next to the Canadian border, to a point 50 miles downstream from the present Tanacross. This also was the initial destination of E.T. Barnette, the founder of Fairbanks. To the casual reader of Alaskan history, the impression has been left that it was he, during his escapade up the Chena River aboard the riverboat *Lavelle Young*, who made the first penetration into the wilderness of the interior of Alaska. Actually, a multitude of explorers preceded him by many years.

Two of these earlier travelers were Arthur Harper, manager of the Alaska Trading Post, and a companion Englishman, Mr. Bates, who came downriver on the Tanana to a point which was close to the mouth of the Chena River. Perhaps one of the greatest deterrents to an aggressive exploration of the Tanana and Yukon Rivers at that time was the rumored belligerence, and subsequent fear, of the indigenous Indians. As it was with the arrival of the initial wave of Pilgrims to Massachusetts, it was only *After* meeting with the white man that the Indians became unfriendly, and *Not Before*!

Despite their fears, by the 1890's the madness for gold had driven these prospectors up the Yukon River to a point where it met the Tanana River. In the meantime, others had come over the mountains from the south and the east to the waters of the upper Yukon to gain entrance to the Tanana River for exploration and exploitation. By 1898, isolated parties of explorers were beginning to cross paths on these routes. In addition, exploratory investigations of two of the tributaries of the Tanana River, the Chena and the

Goodpaster, were carried out. During the summer of 1898, two years after the gold strike of 1896 in the Klondike of Canada's Yukon Territory, Lt. J.C. Castner of the U.S. Army along with several other men, attempted to ascend the Goodpaster River. A shortage of supplies forced them to return to the Tanana River where the local Indians bade them welcome, offering food, shelter, clothing and *friendship*!

After the explorers had regained their strength, the Indians escorted them down the Tanana River to the mouth of the Chena River. At this location they found an old Indian Chief in his fish camp, which was located where the town of *Chenoa* was ultimately erected. This location was named after the river that flowed by the Chief's fish camp, which in the Athabascan Indian tongue means, "River of Rocks". At a later date, the spelling of the town was arbitrarily changed to *Chena*, due to an absence of the letter "O" at the printer's shop. In a few years this town was to play, at least temporarily, a significant role in the development of the Golden Heart of Alaska — Fairbanks!

While Castner's party was in camp with the Indian Chief, he related to them that earlier two river steamboats had gone up the Chena River. These were the *Tanana Chief* and the *Potlach*. Aboard were some eighteen prospectors, with their outfits, who planned to remain there and search for gold during the following winter and summer.

Their stay proved to be futile for they did not find gold. Contrary to the other gold bearing sites in Alaska and the Yukon, there was no surface gold to be found in this area. Instead, what they found was deeply buried, lode gold, which posed totally different personal and engineering problems than those to which they had been accustomed. As best as history can ascertain, these were the first river steamboats to have navigated the Chena River.

Earlier explorations for gold, however, had been carried out. In the 1870's minor deposits of gold were found at the mouths of many rivers along the upper Yukon River. One of these findings was at Birch Creek in 1882. This find turned out to be one of substantial magnitude which lead, ultimately, to the founding of Circle City. This community was located on the south bank of the Yukon River, and became a distribution point for the miners south of this region. However, on the 16th of August, 1896, gold was found in large amounts in the Klondike region of the Yukon Territory of Canada. ***The Gold Rush was on!*** Frenzied people from all over the world cast their shadows on the gold fields, eventually spilling over onto the shores of the Tanana and Yukon Rivers. From several newly-established trading posts along these rivers, miners would venture off, sometimes hundreds of miles, carrying their "outfits" on their backs, in search for the *Valhalla* of golden dreams. One of these men was *Felix Pedro,* an Italian immigrant, trekking south from one of these newly established trading posts, Circle City. In due time, the chance meeting of this miner with Barnette would set off a cataclysmic bonanza the likes of which has had few parallels.

References to the following occur throughout this book: *The Chena Times, Fairbanks Weekly News, Fairbanks Weekly Times, Fairbanks Morning Times, Fairbanks Daily News-Miner, The Frontier Times, Pioneer All Alaska Weekly, The Hot Springs Post and Dawson Daily News.*

The *Frontier Times* was not a bona fide paper, but a promotional publication put out to enhance tourism.

The Chena Times was established in Chena, Alaska, but only one issue was ever printed — July 18, 1905.

The Fairbanks Weekly Times, established September 3, 1905, and *Fairbanks Morning Times*, established May 23, 1906, were published by Times Publishing Company, owned by R.G. Southworth. Located on Front Street, near Kellum, the mailing address was P.O. Box K, Fairbanks City. The company closed November 1, 1916.

The Fairbanks Weekly News, established September, 1903, and the *Fairbanks Evening News*, established May, 1905, were published by J.W. Ward. Located on Garden Island, the mailing address was P.O. Box E, Fairbanks City. The two papers consolidated June, 1907, as *Fairbanks Daily News. Fairbanks Daily News* merged with *The Tanana Miner*, March 9, 1909, to become *Fairbanks Daily News-Miner*. The *Tanana Tribune* was absorbed January 8, 1910 and *Alaska Daily Citizen* was absorbed February 21, 1920. Although it has changed hands several times, *Fairbanks Daily News-Miner* is the oldest continuous newspaper in Alaska.

Pioneer All Alaska Weekly, was established July 3, 1970 , in Fairbanks. It published under this name through October 1, 1992, when the name changed to *All Alaska Weekly*. The publication continues today.

Hot Springs Post, established October 1, 1908, was published in Manley Hot Springs. It was co-edited by George M. Arbuckle and J.W. Ward through November 12, 1908, then edited solely by Arbuckle, until May 15, 1909. Mr. Arbuckle then moved the press to Tanana where he started the *Tanana Leader*.

The Canadian paper, *Dawson Daily News*, is the only non-American paper referenced. It was the first daily newspaper in Dawson, established July, 1899, by owners Richard Roediger and William McIntyre of Tacoma, Washington. It became a weekly paper in 1946, and continued as such until the doors closed March, 1954.

Paragraph 4 references *Steamboats On The Chena*; pg. xiv

Author's Note
Valhalla — A paradise for heroes from Scandinavian mythology

Some of the contributors:

Joe Ashley

Bruce Halderman

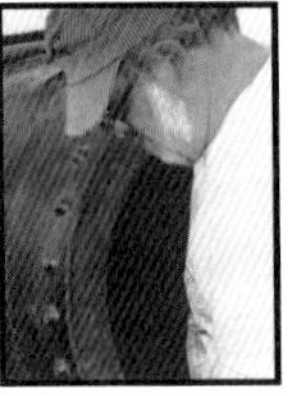

Dan Gullickson

George Simpanen

Olga Steger

Richard "Tark" Tarkianinen

There were three different routes that one could take to reach the Fairbanks mining creeks at the turn of the century — all with their own distinctive character, charm and danger.

The first of these routes was via the village of St. Michael which was located in Norton Sound, an arm of the Bering Sea, just south of the Seward Peninsula. It was perched on a rocky and treeless island just off the coast of the mainland for it was only here that a deep water port existed along the whole northern coast. Passengers and freight were transferred from ocean going vessels onto shallow-draft river steamers. Ultimately, additional transfers were to take place en-route, to even smaller river craft, should developing shallow waters dictate it when traveling to Chena and Fairbanks.

This was an all weather route both for ocean and river travel. Transportation was available from June to October, with the recommended day of departure being no later than the first of June. The journey was started in Seattle, and would take one across the North Pacific Ocean to the Aleutian Islands. Traversing Unimak Pass safely, the vessel would proceed directly north through the Bering Sea to finally arrive in St. Michael. After a transfer to a river steamer was made, the small vessel would proceed southward for 60 miles to gain entrance to the 100 mile wide delta of the Yukon River. Because the mouth of the river had many channels, all of which were very shallow, it was incumbent upon the master of the riverboat to find one suitable to accept his vessel. The shallowness of the various channels were so extreme that a boat could have a draft of no more than three or four feet.

Once the main channel was located, the river steamer pressed along up the Yukon River to Ft. Gibbons (Tanana Village). Swinging off to the northeast at this point, the craft headed up the Tanana River to finally reach its destination either at Chena, or by proceeding further up the Chena River, to Fairbanks. The distance traveled was approximately 4,000 miles at a cost of $125 first class, and $100 second class. The journey from Seattle to Fairbanks took about six weeks!

The second route to Fairbanks required passage across the storm-tossed Gulf of Alaska. This *winter* service originated in Seattle and was limited to the months of November through April. The port of call on this run was Valdez, which is located on Prince William Sound. This trip to Valdez was a treacherous ocean crossing, representing 1,859 miles of extremely dangerous travel. Fare was $45 first class, and $25 second class. From coastal Valdez to interior Fairbanks represented an overland trip of seven to ten days by sleigh during the bitter winter months. The distance traveled was 364 miles at an additional cost of anywhere between $50 to $150, depending on one's taste in travel.

The third alternative route originated both in Seattle, Washington and Vancouver, British Columbia, and traversed the beautiful inside passage to Skagway, Alaska. From there, one would ride the White Pass and Yukon Railway for a 110 miles on their journey to Whitehorse, Yukon Territory. Without

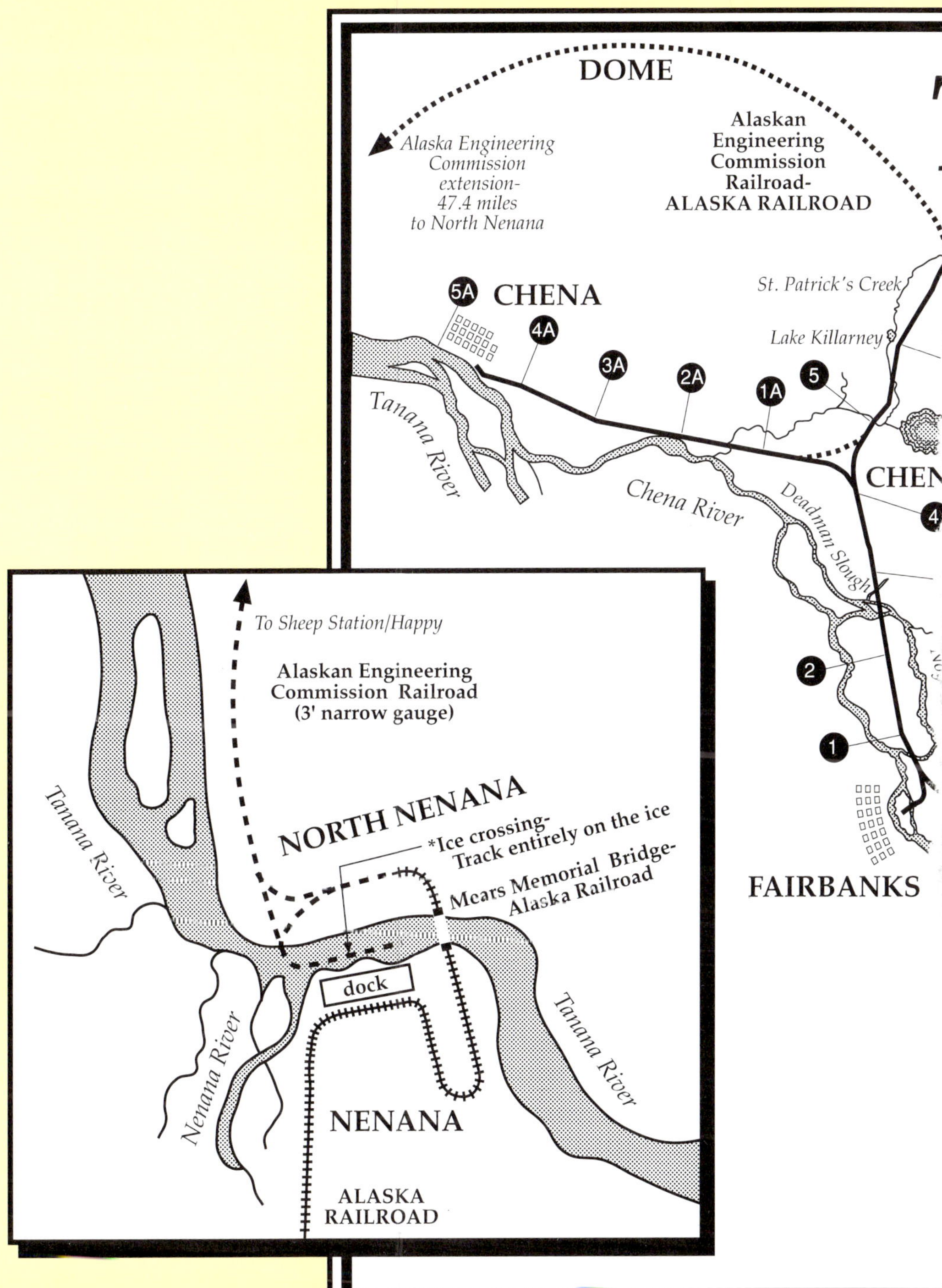
DOME
Alaska Engineering Commission extension- 47.4 miles to North Nenana
Alaskan Engineering Commission Railroad- ALASKA RAILROAD
CHENA
5A
4A
3A
2A
1A
5
St. Patrick's Creek
Lake Killarney
Tanana River
Chena River
Deadman Slough
2
1
FAIRBANKS
To Sheep Station/Happy
Alaskan Engineering Commission Railroad (3' narrow gauge)
NORTH NENANA
*Ice crossing- Track entirely on the ice
Mears Memorial Bridge- Alaska Railroad
Tanana River
dock
Nenana River
Tanana River
NENANA
ALASKA RAILROAD

TANANA VALLEY RAILROAD

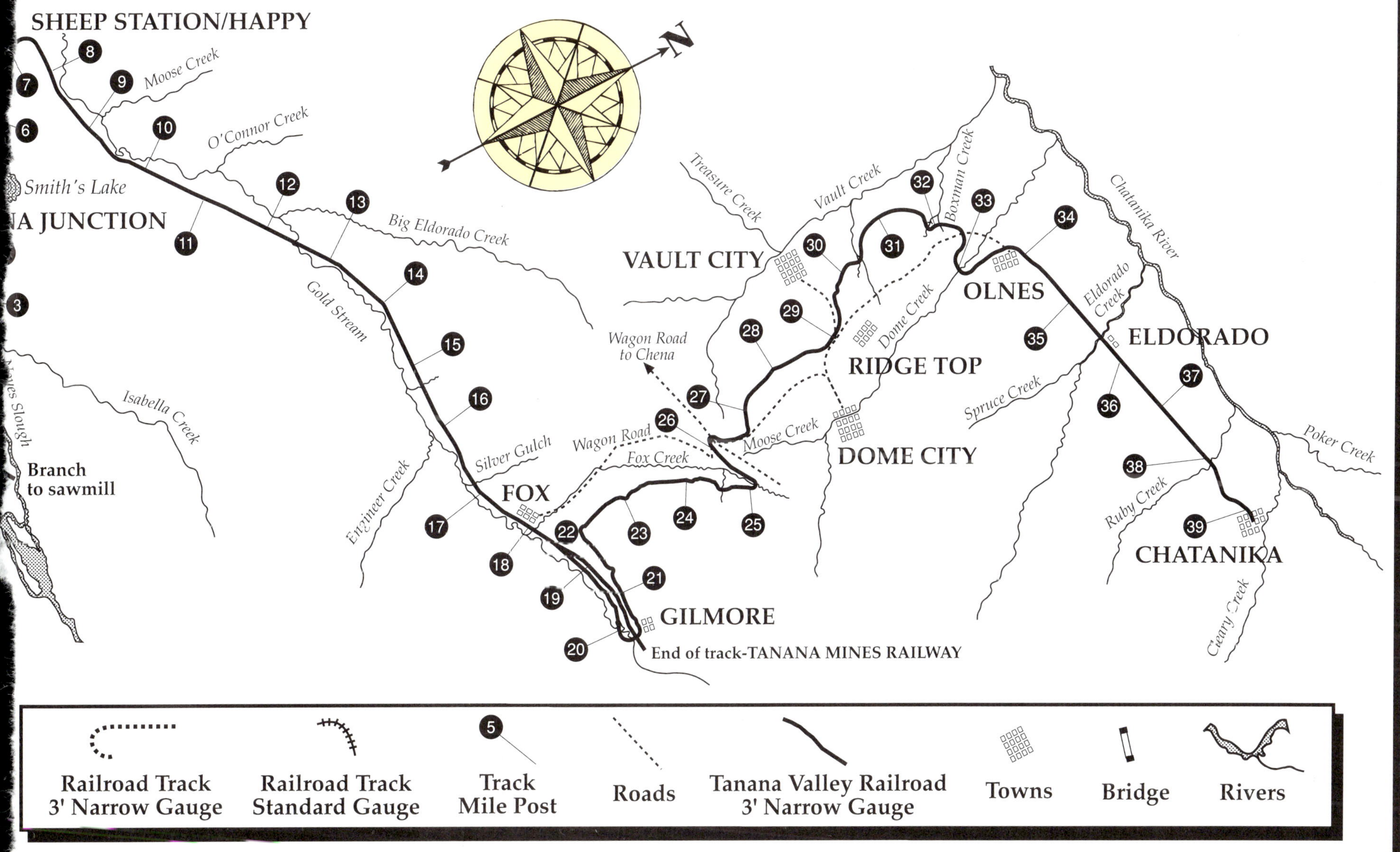

doubt, this railway trip represented one of the standards of the world, by which all others are measured, for mountain scenic grandeur! Once in Whitehorse, transfer was made to one of the numerous riverboats for the journey to Dawson. From there, the trip continued downriver, crossing the Arctic Circle at Ft. Yukon, to the village of Ft. Gibbons (Tanana Village) at the junction of the Tanana and Yukon Rivers. Making a turn of 180 degrees at this point, the river steamer journeyed up the Tanana River to finally dock at Chena and Fairbanks. The total distance travelled was 2,375 miles at a cost of $125 first class, or $100 second class. One of the drawbacks of this route, however, was that passage required an international customs check on at least two occasions.

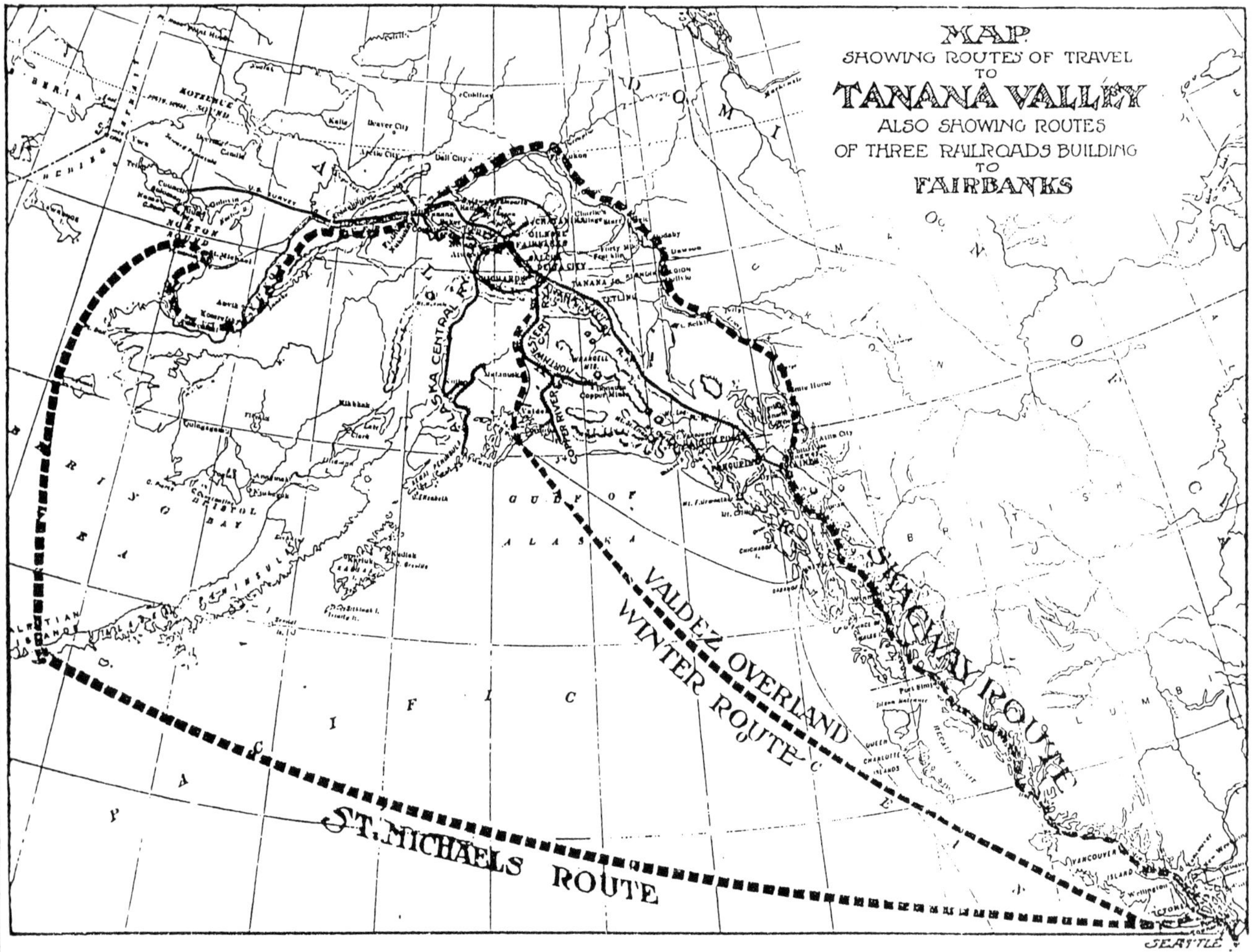

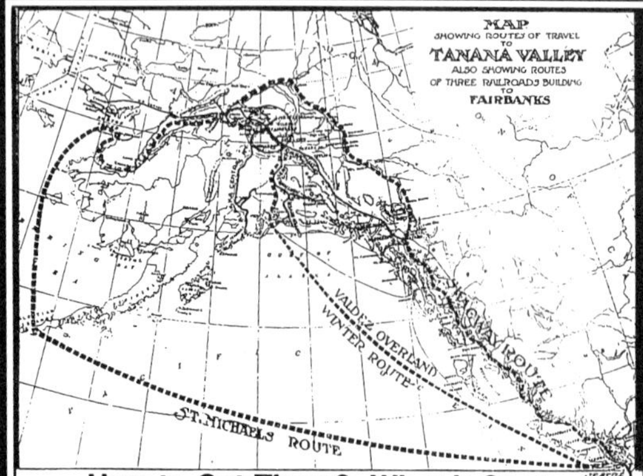

How to Get There? What It Costs.

You can go via Valdez any time between November and April. It is one of the best ways. The fare Seattle to Valdez, including meals and berths, is first class, $45.00; second class, $25.00, and from Valdez to Fairbanks it will cost from $50 to $150, depending on how you travel. A great advantage of this route is that it brings you to the district at the very beginning of the sluicing season, which is harvest time on the gold farms.

The summer routes are open from June to October.

To go by the St. Michaels route you should be prepared to leave Seattle by June 1st. The fare, Seattle to Fairbanks, including meals and berth, is, first class, $125.00; second class, $100.00.

A special very low fare for miners and prospectors will probably be announced by the transportation companies before the June sailing. It will be well to write for these rates.

To go by Skagway you can leave Seattle or Vancouver June 1st. Fare first class $125.00, second class $100.00.

Your ordinary clothing is all you need for summer. A pair of rubber boots or shoes is often useful.

All kinds of clothing adapted to the country can be got at Fairbanks at reasonable prices, so it is not necessary to take any clothing other that what you have.

For information address:

Alaska Steamship Co., Seattle.
White Pass & Yukon Ry. Co., Seattle, or Vancouver B.C.
Northern Commercial Co., Seattle and San Francisco.
Pacific Coast Steamship Co., Seattle.
Schuback & Hamilton, Seattle.

Courtesy of the Candy Waugaman Collection.
Printed by the Fairbanks Daily News-Miner, June 9, 1991.
Originally printed in 1907 by the Fairbanks Chamber of Commerce.

MAP
SHOWING ROUTES OF TRAVEL
TO
TANANA VALLEY
ALSO SHOWING ROUTES
OF THREE RAILROADS BUILDING
TO
FAIRBANKS

VALDEZ OVERLAND WINTER ROUTE
SKAGWAY ROUTE
ST. MICHAELS ROUTE
SEATTLE

CHAPTER 2

FOUNDING OF FAIRBANKS AND CHENA

Map of the Tanana Mines Railway, Tanana Valley Railroad and the Alaskan Engineering Commission railroad. Insert depicts the winter ice crossing between North Nenana and Nenana.

Initially, it was the explorers and adventurers, and not the prospectors who made the trip to the interior of Alaska. Riding on the waves of humanity that were sweeping into the Klondike and subsequently into Alaska was E.T. Barnette, a man of questionable character, whose pot of gold was to come from the dangerous and back breaking work of others. Earlier he had corresponded with John Jerome Healy, a man of equally dubious character, relative to a possible business adventure, that ironically was to become a hallmark in the history of interior Alaska.

What Healy had envisioned in his correspondence with Barnette was a system of railroads in Alaska which would not only criss-cross the territory, but would be extended to Siberia by crossing the Bering Sea. In effect, what he was proposing was a land passage, by railroad, from Paris, France, to extend across Europe, Asia, and then North America via Alaska, ultimately reaching New York after crossing three Continents.

Healy's most immediate concept, however, was the one that really excited Barnette's imagination, that is, the construction of a railroad line from the ice free port of Valdez, on Prince William Sound, to Eagle. This, he reasoned, would be the shortest route to the Klondike gold fields. Further, it would be an "All American" route. None of these dreams was ever full-filled, but it did launch Barnette on his dream to build a trading post on a site which represented the halfway point on the projected railroad from Valdez to Eagle. It was called Tanana Crossing. Today it is known as *Tanacross*. This development never did occur, but its pursuit lead to a far greater development — the founding of *Fairbanks*!

E.T. Barnette, pg. 18

Traveling to Alaska's interior via St. Michael, Barnette, like Columbus who was seeking a shorter route to India, went in pursuit of his own short-cut to Tanana Crossing (Tanacross). That site was located quite a distance upstream on the Tanana River from the site which was to eventually become Fairbanks. To reach it at that time was an impossible task, for one had to pass through the formidable and dangerous white water, which existed at that time, known as the *Bates Rapids*!

Author's Historical Note
Bates Rapids — Extended west from Big Delta on the Tanana River — 69° 14' N, 146° 00' W. Named by Lt. H. T. Allen in 1887 for Mr. Bates, an Englishman, who is reported to have descended the Tanana River.

Since Barnette's destination on the upper Tanana River was at least four hundred miles beyond any point thus far reached by any river captain, his requested passage at St. Michael, for himself and 135 tons of cargo, was met with ridicule and with total rejection. All of his hesitancy was further compounded by the fact that in a trial cruise about the harbor at St. Michael, his initial craft, the Arctic Boy, hit some submerged reefs and sank. It was a total loss.

The hesitation on the part of the river captains to take Barnette to his destination was due not only to the hundreds of miles of shallow water with all its insidious sandbars, that their river crafts must ply, but also because of the un-navigable Bates Rapids which had to be traversed to reach his destination. Finally — the captain of the *Lavelle Young*, a river steamer with a three foot draft, agreed to take him to his destination, primed by the appropriate financial stimulation, but with the following stipulation: The skipper, Charles Adams, would charge so much per ton for the trip up to the Chena Slough, and then there would be an additional payment per ton of cargo carried beyond the slough up through Bates Rapids to Tanacross. If it became evident that the craft could not navigate beyond the Chena Slough, Barnette, his entourage, and his supplies would be put off the boat *wherever they happened to be*!

On August 26, 1901, E.T. Barnette was told to "Get Off" the *Lavelle Young* at approximately this same location. Next to the *Lavelle Young* is the riverboat *Isabelle*. The *Isabelle* was purchased in Seattle and shipped piecemeal to St. Michael. It was re-assembled there in the Northern Navigation Company ship yards, utilizing the machinery from the sunken *Arctic Boy*. *Isabelle's* last known recorded landing in Fairbanks and Chena was in 1906. Landing is along the south shore of the Chena River, Fairbanks, Alaska, July 23, 1904. Photo credit: University of Alaska Fairbanks, Alaska and Polar Regions Department, Acc. #73-79-91N, in the archives, of the T. Troseth Photograph Collection.

On August 7, 1901, the steamer pulled up anchor at St. Michael and proceeded up the Yukon and Tanana Rivers — a trip that was to be uneventful until the Chena Slough was reached. Further passage beyond this point proved to be futile. Facing the prospect of being forced to evacuate the steamer as was stipulated in the contract, Barnette was able to convince the captain to make one more attempt, but this time up the Chena Slough. According to the information that he had received from some of the Indians, the slough could be used as a by-pass around the white waters of Bates Rapids. At the turn of the century, the north end of the slough, *was open* and allowed the Tanana River to send some of its flow down into the Chena River — circumventing the dreaded rapids. In later years, the rapids would disappear, and the north mouth of this channel would also close off to form, with the Little Chena River, what is known today as the Chena River.

After expending much effort, tribulation and frustration attempting to steam up the Chena, Barnette's realization of inevitable failure became a fact. True to his contract, the captain reversed his course and proceeded downstream to a point about six miles from the confluence of the Chena and Tanana Rivers. Unceremoniously the gangplank was lowered onto the wooded south bank of the Chena River. And at about four o'clock in the afternoon on the 26th of August, 1901, Barnette and his band of voyageurs was told to "GET OFF"! Little did they realize that this was the first in a *Trilogy* of events that was to lead to the founding of an academic, mining and cultural community in the heart of Alaska — *Fairbanks*!

Paragraphs 2 & 3 reference
E.T. Barnette, pgs. 23 & 24

While all this activity was unfolding down on the south bank of the Chena River, high up in the adjacent hills north of this exact location, at this very same time, Felix Pedro, the Italian immigrant and miner, was tramping his way south from Circle City, a town that existed primarily as a distribution center for the miners and as an oasis from the wilderness.

Felix Pedro saw the smoke coming from the riverboat *Lavelle Young* as it hesitantly attempted to proceed up the Chena River. He became overjoyed with the prospect that an encounter with this boat might give him the opportunity to buy badly needed supplies so that he might continue on with his mining labors. This way he felt that he could save 330 miles on a round trip to Circle City to replenish his larder, fighting wild animals and mosquitoes all the way.

What happened next, as he emerged from the woods, can only be a conjecture. However, it would be reasonable to assume that he, in a rush of Latin machismo, opened his out-stretched arms to Barnette and his party and greeted them with, "Hey Compagno — Buon Giorno." Unbelievably, this occurred just a short time after Barnette was told to get off the boat. Within one year after this chance meeting, Felix Pedro would make two gold strikes out in the hills — one only 12 miles from town! These strikes occurred on July 22, 1902, with the second one following on September 11, of the same year. And, with these strikes, the *Gold Rush* was on! Shortly thereafter, claims were made at Fairbanks Creek, Little Eldorado, Dome Creek, Vault Creek and at many other sites. Thus the need was established for a transportation system that could haul heavy tonnage — a railroad! The second chapter of the trilogy was now completed.

At the same time that the first two acts of this trilogy were being played out, the third and final episode had long been in production. Early in 1901, George Belt and Nathan Hendricks, agents for the North American Transportation and Trading Company (N.A.T. & T. Co.), established a trading camp on the south shore of the Tanana River opposite the mouth of the Chena River. However, as early as 1898, much before Barnette's arrival on the scene, this same company cached 70 tons of supplies and building materials on the *north* bank of the Tanana River at the junction where the Chena River flowed into it. This cargo was apparently deposited by the riverboat *Tanana Chief* on the same location that had once served as a fish camp for the Indian Chief who told of earlier explorers going upstream on this very same boat.

Steamboats On The Chena, pg. 23

By 1902, Belt and Hendricks had decided to relocate their trading post from the south bank of the Tanana to the north shore at its confluence with the Chena River. Several factors may have been responsible for this move, but certainly anticipation, as witnessed by the early drop-off of supplies at this new location, must have been an impetus. While trade with the Indians was always the prime reason for the existence of the trading post, the emergence of multiple bonanza gold strikes, bringing a continuous flow of prospectors and miners heading up the Chena River, forced Belt and Hendricks to re-think their priorities. Further, the United States Government was stringing a telegraph line from Valdez to Eagle, and then on to Dawson in the Yukon Territory. In addition, a secondary line would come down along the Tanana River, starting at Tanacross. This line, on its way to St. Michael, would pass right by the new landing site at the Tanana-Chena River Junction. Since most of the travelers to the

Interior seemed to be passing through this site, it appeared to the entrepreneur that this would be a very logical place to create a center for commerce, transportation and business. Thus were sown the seeds which ultimately germinated into the town of *Chena* — and, ironically enough, the seeds for its demise!

Chena's early history is really a tale of two cities, that of itself and of its upriver rival, Fairbanks. Chena was first called Tanana City. Later, it had its name changed to *Chenoa*, which was ironically the initial name given to the present city of Fairbanks. The crucible of their rivalry was transportation and *Gold*! It wasn't that either town had gold strikes in their community, but rather, who was in a better position to serve the miners in their business and civic affairs.

Since the primary mode of transportation into the Interior during the early nineteen hundreds was river travel, Chena — which sat right on the shore of the principal waterway of commerce, the navigable Tanana River, felt that it should become the transportation hub of the North. This concept was strongly endorsed by Belt and Hendricks who felt that fluctuating water levels, primarily on the Chena River, would preclude any development six miles further upstream at Barnette's Cache which would later become the city of Fairbanks. Shipping season on the Chena River was at the mercy of variable water levels because of seasonal and climatic conditions. Boats with drafts of more than two feet would be limited or prevented all together from coming up the Chena. Therefore, any possible sustained commerce would become erratic or completely choked off.

At approximately the same time in 1902 that plans were being formulated for the development and construction of the town of Chena, fate was driving the first nail into the coffin for the ultimate demise of this community. Felix Pedro, the Italian immigrant who greeted E.T. Barnette at the time of his crass ejection from the *Lavelle Young* during the latter part of August of 1901, returned once again to this newly formed trading post known as Barnette's Cache (Fairbanks). Instead of being a beggar for supplies, he returned as potentially one of the richest men in the region. He *struck it* — ***Gold***! Letting his find be known initially only to friends, it was only a matter of time until the news would be leaked out, and the whole region would become covered with claims.

Barnette, who had been attending business matters in the states, returned to his cache to find a mini-stampede going on. All plans for a trading post upriver at Tanacross were now abandoned. Not only did he firm up his determination to stay at the present location, but true to his emerging character, he elected to exploit this strike in order to bring more miners into this region — and to his store! In order to inflame the minds and passions of the miners in the Klondike camps, he sent out his Japanese cook,

Author's Historical Footnote

Getting a stuck riverboat free was a major undertaking and wasn't taken lightly by boatmen. If the boat was stuck on a newly formed and loosely packed sandbar, it was sometimes possible to just lighten the load and back the boat hard to pull off. But usually this wasn't the case. "Lining" was the term used for trying to winch a boat free. "Sparring" was the most commonly used method. Two large spars or "legs" were bolted to the boat's superstructure and set in the river bottom by derricks on the deck. Moving slowly, the captain would try to transfer much of the weight of the boat onto the legs and so leap ahead. At the end of each leap, the spars had to be reset and this action continued until the boat was free. If a combined effort of unloading, lining and sparring didn't work, the boat was forced to hope for another to come along and help pull it free. If winter came first, the boat was usually considered a total loss and abandoned, because few boats survived freeze-up and breakup in the rivers. The *Gold Star* was lost 1900, 15 miles below Chena, the *Rock Island #1* was lost just above Chena in 1906, the *Reliance* in the Tanana near Chena in 1917 and the *Tanana* was lost near Chena in 1921.

Jujiro Wada, to the Yukon. With him he brought great "tales" of the unlimited gold which had been found in the interior of Alaska.

Wada left Barnette's Cache three days after Christmas of 1902. Nineteen days later, he was in Dawson City extolling the news of the latest bonanza to have hit the north. Within days of his pronouncement, the stampede was on! Across the hills and valleys, during this early January of 1903, came hordes of men. Mining "outfits" were strapped to their backs as they forced their way through the ice laden snows to the point of that fortunate landing on the Chena River — Barnette's Cache!

E.T. Barnette, pgs. 42 - 44

Prior to this series of unfolding events, another chance meeting occurred which would change the whole competitive scenario that was beginning to develop between the emerging towns of Chena and Barnette's Cache (Fairbanks). This occurred during the summer 1902 when Barnette was once again in St. Michael making preparations for his second trip up the Chena River. Felix Pedro was as yet to make his resounding second gold strike. It was also six months prior to the stampede from the Klondike which Barnette had orchestrated.

Author's Historical Note

Such was the demand for firewood, that whole hillsides and entire valleys were clear-cut, and even the brush piles one normally sees in clear-cut areas were carted off and utilized for cooking and heating fires. Devastation was near total, and these clear-cut areas repelled rather than attracted settlers. Ironically, environmental damage (runoff) from these clear-cut areas piled more silt into the riverbeds and creeks, drying many up completely causing myriad problems for both the miners and the boatmen, and reducing the numbers of wild-game and fish in the area.

In time, Barnette had secured a new boat for this trip and named it *Isabella* after his wife. While making preparations for departure, he met James Wickersham who, through his control of the Third Judicial District of Alaska, represented the real power in this part of the country. Meeting Barnette on a secluded beach in St. Michael, Wikersham asked Barnette if he would name his cache on the Chena River after Charles W. Fairbanks, Senior Senator from Indiana. Wickersham then promised to do everything in his power to see to it that Barnette's Cache would become a viable entity. Barnette could see the handwriting on the wall and quickly acquiesced to this request.

On September 8, 1902, a short while after Barnette returned to his trading post, a meeting of the miners was held at Pedro Creek, not too far from Barnette's Cache, to develop a new political entity for the immediate region. Barnette was present at this meeting and was made Temporary Recorder of the district by the miners. During the course of their deliberations, Barnette presented to them the option of naming this newly developed area —*Fairbanks*! This they agreed to do, and while it was nothing more than a political *Quid Pro Quo*, Wickersham, true to his word, exercised his prerogative by propelling Fairbanks into becoming the most prominent community in Alaska. By the same token, it spelled the demise for the town of Chena, even before it was built.

E.T. Barnette, pgs. 29, 30 & 38

During the year of 1903, miners by the hundreds and ultimately thousands, descended into the Tanana Valley. Part of this mass exodus into the area was in answer to the false gold rumors which were spread, in the Klondike, by Barnette. By and large it was the miners who came to Fairbanks, while the entrepreneurs set up shop in Chena. These miners were seasoned veterans from the Klondike and Nome, experienced in the stresses and problems of the sub-arctic, and were prepared to go to work straight-away.

Sadly, the Indians, the indigenous people of Alaska, were **prohibited by law** from any mining! They were, unfortunately, limited to only the menial jobs in the camps.

Much to the dismay of the miners, gold just for the digging, as was experienced in the Klondike and Nome, was there all right, but not in shallow diggings as they earlier had experienced. Here, one had to thaw and dig, thaw and dig, and continue to thaw and dig farther down to levels as much as 316 feet below the surface. This was not the land of the pick and shovel alone; it was the land of steam boilers, winches, hoists and energy output as supplied by *unlimited wood and water!*

As the only source of energy, wood became increasingly scarce as the years went by. Water was always a crap-shoot, for it came from shallow creeks which were totally dependent on the climatic conditions. It soon became evident to the miners that the riches of the region were not to be taken for granted. Early arrivals had laid claims to all the land and it became increasingly evident that, at least on the topsoil, there was no gold to be found. Disillusioned and feeling betrayed by the false reports of the bonanza as was initially perpetrated by Barnette, the miners began to leave the region with extreme bitterness! While both Chena and Fairbanks suffered from this population decline, Chena took the brunt of it.

Initially, it was Chena's position on the Tanana River that made it, with some justification, the premier community in the Interior; and as such, it had the most to lose. Chena sat on the north bank of the broad Tanana River, hindered in navigation by changing water levels only to a minor degree. As such, the initial influx of the newly arrived people felt that this was the place to settle. Like locusts in the middle of a corn field, they swooped into town, staking claims with such total disregard for other people's property that whole cabins were stolen, in their entirety, right off of other settler's claims. By the end of 1903, which was still within the immediate time frame of the origins of the city, Chena could record only *23 residents*. To further compound the misery of the forlorn lot, a shortage of food developed resulting in near-starvation conditions. Wild game became a vital necessity for subsistence and survival. Even the scrounging dogs began to look good! As a result, normal accepted civility between the residents of these two neighboring towns degenerated to the point that merchants in Chena would not sell supplies to residents of Fairbanks.

While all this pandemonium was going on in Chena, Judge James Wickersham arrived in Fairbanks for the first time, on April 9, 1903. True to the Russian adage, "Everything in life is politics," he put into effect the very political decisions that insured the survival of Fairbanks. Acknowledging the fact that

WATER GROWING SCARCE NOW

Operators On the Creeks Must Have Rain Soon.

Rain is very badly needed on the creeks, where work is being affected some by scarcity of water. On Pedro creek Julius Gius, Harry Atwood and Mrs. Pedro are the only concern of large size sluicing now, the rest being unable to do more than to get ready and await more water.

The same conditions prevail on the other side of the divide. Both Fairbanks and Ester creeks are almost dry at the present time, heading on very low divides, and Cleary creek is also seriously affected.

Fairbanks Daily News-Miner, June 5, 1915.

E.T. Barnette, pgs. 50 & 82

The biggest "thief" in Fairbanks during the gold rush days was "Mr. Bruin", himself. So thirsty was he that he "shared" much of the stolen milk with his fur coat. Photo credit: University of Alaska Fairbanks, Alaska and Polar Regions Department, in the archives, of the Howard Henry Collection.

Chena was a more desirable port city, he never-the-less tilted his hand toward Fairbanks, the town that he was instrumental in naming. Through his political machinations, two important transfers were made from Eagle to Fairbanks. First to be transferred in May, 1903 were the offices of the Third Judicial District. Next to follow, later in the summer of that year, was the Mining Claim Recording Office. In the meantime, during the same month, a postmaster was commissioned and a courthouse and jail site were established. To further insure this community's destiny as a leading city in Alaska, he made a "special request" to the United States Army Signal Corps as they were stringing their telegraph cable from Tanacross to St. Michael via the town site of Chena. In accordance with his wishes, Fairbanks became linked to this vast telegraph network during the month of July, 1903. On November 10, 1903, Fairbanks had reached such a degree of development that it was incorporated as a city. Thus, in the space of two years, this settlement evolved from a wilderness landing in 1901 to become the very heart of Alaska and the gold mining center of North America!

Steamboats On The Chena, pg. 21

Concurrent with this exodus from both Chena and Fairbanks, however, was a series of new strikes that occurred during the autumn of 1903. Out in Fairbanks Creek, gold was found in relatively shallow ground where bedrock was only ten feet below the surface. Shortly thereafter, strikes on both Cleary and Ester Creeks were made. These proved to be the ultimate discoveries, for between the three strikes two-thirds of the total gold mined in the Tanana Valley was to be recovered by 1910! This amounted to more than 30 million dollars worth.

E.T. Barnette, pgs. 63 & 67

Paradoxically, with the series of strikes that occurred at this time, a scene of mass confusion developed bordering on bedlam and schizophrenia at the same time. On one hand were ship-loads of disgruntled and discouraged miners heading back to Dawson; while at the very same time, other ship-loads of miners were on their way, in a state of panicked glee, to the Tanana!

However, this new batch of miners were a different lot of people from the original ones, for they had a more realistic appraisal of the problems to be encountered. Martin Harrais, route developer of the *Tanana Mines Railway*, a leading proponent in the development of Chena, and a man who was about to enter into the folklore of this new area, stated it most succinctly:

> "The stampeders brought their mining plants from both Dawson and Nome. They proceeded to look the creeks over, get a lay on the ground, put their plants on it, *and go to work.* The development was rapid, and by mid-summer the miners were taking out gold by the ton."

The Way It Was, pg. 139

By the spring and summer of 1904, the Fairbanks stampede was in full swing. It was a repetition of the Dawson and Nome days, except that there were, generally, no Cheechakoes among them. The few Cheechakoes that did come on the scene invariably came up from the outside and were quickly absorbed into the flock so that they went un-noticed!

Arthor's Note:
Cheechakoes — Newcomer to Alaska or the Yukon

While a prospector could make his claim anywhere in unaccountable land, the need for adequate transportation limited his movement into the wilderness. As a result, during the years of 1903 and 1904 a crude system of roads developed, radiating out from Fairbanks to the drift mines scattered across the hills. It was soon recognized that the cost of moving freight by a horse-drawn team from the docks at Fairbanks or Chena, if it could be done at all, was financially prohibitive. Concomitantly, while team transportation was a vital asset to the new towns and distant mines, this form of restrictive movement became an obstacle to the growth and development of the region. Horse-drawn teams were, in themselves, limited to what they could haul. Six straining animals, climbing the hills which led to the drift mines were no match for the heavy boilers which were indispensable to the mining industry.

Paragraph 1 references an article written by Falcon Joslin and published in *Alaska-Yukon Magazine*, pg. 247, 1908

Even when trucks came on the scene their loads were not only limited, but completely shut down, for all practical purposes, in the winter. It was not primarily the snow, cold and winds which bogged down the trucks, but rather the solid rubber tires that were used during this era. With the onset of bitter, winter cold weather, these tires would become very hard and brittle, ultimately losing their ability to develop traction. It was sometime later that the pneumatic tires came into use which would allow some winter transport, but could never meet the needs of the mining community. Further, the local roads became bottomless quagmires with the onset of spring breakup. The process of freezing and subsequent thawing of the ground would fracture the surface of the exposed trail to such a degree that spring rains made short work of transforming the once solid base into a spongy morass supporting only that which could float. During these periods, freight was pretty much limited to what pack animals could carry on their backs for distances of 20 or 25 miles from the steamboat landings at Fairbanks or Chena. Winter, in spite of snow and cold, paradoxically provided a solid base whereby heavy loads could be transported by sled out to the camps. However, the load was still limited to the pulling power of six horses.

Paragraph 2 references an article written by Judge James Wickersham and published in *Old Yukon*, pgs. 474-475, 1938

Confronted by high teamster rates and roads which simply served as convenient moose trails, the need for a reliable transportation system became paramount for the survival of the mining industry and for the newly emerging community of Fairbanks. Existing roads were clogged with snow in the winter time, convoluted and fractured by the spring thaws, and submerged in a sea of mud or washed away altogether during the summer rains. It was becoming increasingly evident that an all weather transportation system was needed, capable of hauling unlimited tonnage, at a reasonable cost, if the mining operations and communities were to survive. The criteria set forth represented the genesis of the only "tool" that could measure up to these requirements — a ***Railroad***!

CHAPTER 3

Construction of the Tanana Mines Railway

Newly laid track with unfinished ties and no tie plates. No locomotive is present on the construction site.
Photo courtesy of the Paul Solka Collection

Porter No. 1, 0-4-0 a saddle tank locomotive with its crew and visitor. Location unknown.
Postcard courtesy of the Candy Waugaman Collection

During this period of mayhem and bitterness in the Tanana Camp, another set of players were to arrive on the scene in 1904 from the Klondike. They were entrepreneurs who had done well in the Yukon, and had now begun to take great interest in what was happening in these newly emerging gold creeks. One of these people was *Falcon Joslin*, an attorney who had been involved in numerous business ventures in Dawson City. Another individual was *Martin Harrais*, who along with Frank Smith, and John Joslin, Falcon's brother, were to play key roles in the development and construction of the ***Tanana Mines Railway***. Two years later, this line was to become the nucleus of the larger ***Tanana Valley Railroad***!

Falcon Joslin was born in Belleview, Tennessee on September 27, 1866, the son of John M. and Margery Joslin. He graduated from Vanderbilt University, Nashville, Tennesse with a B.L. (Bachelor of Law) degree. Initially, he practiced Law in Seattle, Washington, but by 1897 he too, followed the mass migration of gold prospectors north to Dawson City and the Klondike of Canada's Yukon Territory. A few years after his arrival, in May of 1900, he married Lora Price. During this same period of 1897 to 1902, he organized the Dawson Electric Light and Power Company and served subsequently as its president. Seeing the great need for energy that was vital to the mining interests in the Yukon, he promoted and helped build the Coal Creek Coal Company, an enterprise designed to harvest the prevailing seems of local coal. A three foot narrow gauge railway was laid a distance of twelve miles from the coal deposits at Coal Creek to a point along the Yukon River which was some six miles below the Fortymile country along this waterway.

ery truly yours,

Signature of Falcon Joslin
Courtesy of the Daniel Gullickson Collection

This last adventure in the Yukon Territory was of great significance, for when Falcon Joslin left for the greatest of his developmental projects, in Alaska, he did not go empty handed. In time, one of the locomotives from this railway, a Porter 0-4-0 saddle tank, was to follow. It would be the first locomotive on the Tanana Mines Railway and, very appropriately, was designated as engine No. 1. Further, it was destined to bridge three centuries, for it was constructed in the 19th century, labored in the 20th century, and will survive into the 21st century as a living history of the past.

Coal from these mines was used to fire the boilers of the Dawson Electric and Power Company and those of the Canadian river steamers that plied the Yukon River. In some respects, this coal was nearly as valuable as the gold taken from the ground, for it was the "stored energy" that propelled the Klondike mining adventure. By the same token, its absence from the mining scene in Alaska caused untold hardships and uncertainties in the mining camps.

Locomotive No. 1 is currently being restored to its original form by the friends of the Tanana Valley Railroad. It is scheduled to go into operation in Fairbanks during the summer of 1999.
Photo courtesy of the Richard "Tark" Tarkiainen Collection

Although Falcon Joslin was to assume other prestigious positions — such as President of the Alaska Wireless Co., in 1910, and President of the St. Elias Oil Co., of Katalla, Alaska in 1916 — his crown jewel was the role that he played as promoter and President of the Tanana Mines Railway. In 1907, this line was further extended from its original terminus at Gilmore to become the Tanana Valley Railroad.

Paragraph 1 & 2
The Way It Was, pg. 139

Tanana Valley Railroad Bond
Photo credit: University of Alaska Fairbanks, Alaska and Polar Regions Department Acc. #92-008-01, in the archives, of the Tanana Valley Railroad Collection

However, the initial on-the-spot supervision of the construction of the railway was not carried out by Joslin, but rather by his 38 year old friend, Martin Harrais. He, too, was in the Klondike as a entrepreneur and miner with Falcon Joslin, where he made millions of dollars. After working his claims on the *Bonanza* and *All Gold* creeks, Harrais left Dawson during the autumn of 1903 to travel 1,000 miles by dog team, to the Tanana Valley.

Immediately upon his arrival in the Tanana Valley, Martin Harrais, along with fellow entrepreneurs Frank Smith and John Joslin, set about the task of determining the route for the new railway. Using a *brown paper bag* to record their engineering concepts, these men plotted the survey route for this line. The pike was to run from the developing townsite of Chena, across the lowlands northeast of town, and then through the muskeg-choked Goldstream Valley to Fox — ultimately reaching Gilmore which was 21 miles from Chena. Finishing this survey in the dead of winter, Frank Smith wrapped it up, put it in his pocket, and headed back to report to Falcon Joslin, who was still in Dawson City.

Paragraph 3 was compiled with references to U.S. Government, Congressional Floor Reports: U.S. House 1908:12; U.S. House 1913:44 and U. S. Senate 1909:247 presented by Joslin; U.S. Senate 1912:7-9, and a letter written by Al. George to Land Management Officer/Files, and T.V.R.R. Annual Report, 1908

After collecting all the data necessary to present a formidable argument for the construction of the railway, Falcon Joslin went off to London, England to secure financial backing for his project. He approached *Close Brothers and Company*, the financial backers of the *White Pass* and *Yukon Railway*. He was successful, and secured $400,000 with promises of more money should the railway be extended. With this capital from the White Pass interests, under cover of an English umbrella (Close Brothers), 3,000 shares of preferred stock and 10,000 shares of common stock were issued. The financial aspects thus settled, the *Tanana Mines Railway* was born, incorporated in 1904 in the State of Washington. However, when the time came for the extension of the railway, the money that originally was promised by Close Brothers to Falcon Joslin for this specific purpose, would not be forth-coming! Close Brothers argument was that they were financially heavily involved with the *Copper River* and *Northwestern Railroad* being built north out of Cordova. In actual fact, they wanted to withdraw altogether from their involvement with the Tanana Mines Railway. This was to be the case in 1907 when the railway restructured its financial base.

Acting almost in unison, Martin Harrais and Falcon Joslin went about their separate tasks during the late fall and winter of 1904, directing the operations needed to get the railway built and in operation as soon as possible. While Joslin was making his way in from Dawson City to the new Fairbanks, across muskeg and mountains laden with snow and frozen in time by severe cold weather, Martin Harrais, ever the entrepreneur, turned his aspirations and developmental skills towards Chena. Not wanting to remain idle during the winter, and sensing the many needs of the miners which were to follow, he enveloped himself in a montage of developmental activities. His belief on the future of Chena as a viable community was so great that he invested $86,000 of his own money in its development.

He built a saw mill to meet the immediate construction needs of the town and of the miners by scrounging around for derelict parts from abandoned riverboats, from which he was able to construct a power unit to cut logs into slabs of wood. Still using his own money, he built the Chena Lumber, Light and Power Company and the largest shipyard in the North! In time he was to meet and marry a local school teacher. Later, he dabbled in politics and was made a delegate to the Democratic Convention in Sitka.

Thanks to his efforts, Chena, for the second time, started to grow. Streets were laid out in a grid bearing names similar to that of many of the states. This was *not* the case. Rather they were actually named after the battleships that served in the Spanish-American war of 1898 — which were named after the various states. Chena developed a public school system, telephone system, hospital, jail, a 500 seat public amusement hall, several newspapers, and a police and fire department. However, the primary reason for Chena's existence was the railroad and its shops, for long after the town ceased to be a viable community, the railroad was still on the scene. Even during the early stages of the town's construction in 1904-1905, there was some conjecture that the decision to use Chena as a terminal could have been influenced by the heavy financial investment that was being made to the town by Martin Harrais. Already, there was speculation that although one could rely on the Tanana River for navigation, the uncertainties associated with navigation on the Chena probably did not represent too much of an obstacle. Time ultimately proved that this was a fact. As a result, Chena's heyday was short lived. This occurred during the years of 1904-1905, which ironically were the very years that it was under construction. Population at this time was 450 souls, which was almost as large as the developing Fairbanks.

While Martin Harrais was involved with the development and construction of the townsite of Chena, Falcon Joslin came from Dawson City with the intent of building at least ten miles of track before freeze-up. To keep the cost of construction down to a minimum, he elected to build the railway to a gauge of three feet (the distance between the rails). This is in sharp contrast to the gauge used by all North American railroads, which is 4 feet, 8½ inches between the rails. Heavy bridges and extensive track work would not be needed to support the much smaller locomotives and rolling stock. Curves could be sharper and thus extensive excavation, especially in the hills, would not be necessary. These considerations led to an increased speed in construction, and also substantial financial savings.

Joslin had an immediate goal and an ultimate plan. Initially he wanted *to build the line from Chena to Fairbanks* and then, from a junction point along this line, an extension out to the mining creeks in and around Gilmore. Further, he envisioned an extension of the line out to Chatanika, which became a reality, and a myriad of other lines out to Nome, to Rampart and to Haines in southeastern Alaska, for a total of 1,400 miles. These latter routes did not become realities.

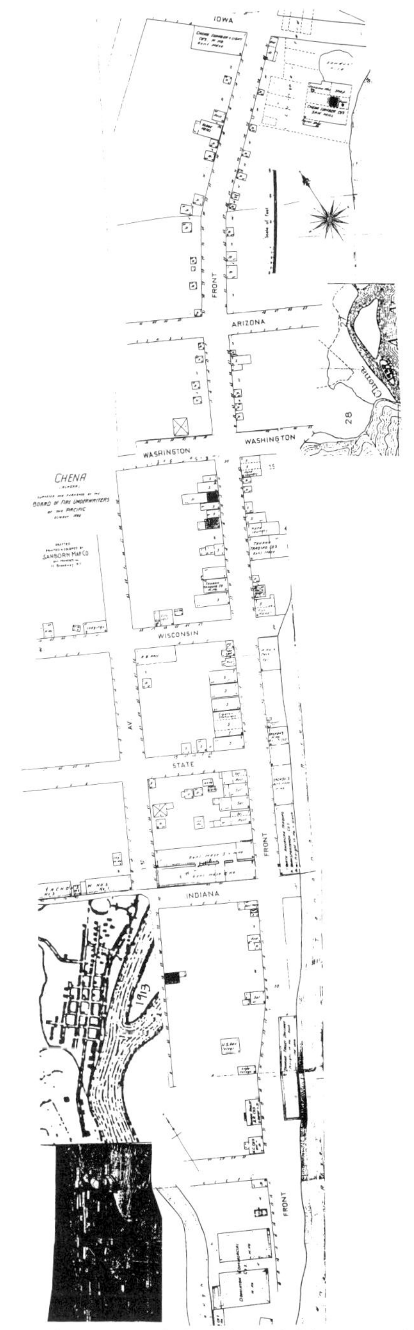

Survey plat map of the city of Chena, 1908.
Courtesy of the Paolo Greer Collection

Paragraph 4, page 25 references an article written by Falcon Joslin and published in *Alaska-Yukon Magazine*, under the title "Railroad Building in Alaska". Pgs. 245 & 246, January, 1909

Paragraph 1 consolidates references Dawson Daily News, November 18, 1904 and *The Way It Was*, Pg. 140

Dawson Daily News, August 31, 1904

Construction, at least in theory, of the Tanana Mines Railway, began during the autumn of 1904. It was, at best, a futile attempt, for the ground soon became frozen. Further diggings had to be terminated because hitting the ground with a pick was just like striking a rock from which only bits and pieces could be dislodged. However, the movement of construction materials and rails into Chena started in earnest about this time. There is, however, some conjecture as to when and where the first consignment of freight for the Tanana Mines Railway took place. One scenario is that Falcon Joslin made early arrangements for the deposition of workers supplies, tents, wagons, scrapers, wheelbarrows, and hay and grain for the horses at the time of his departure for Chena. It is likely that these supplies came up the Yukon River from St. Michael ultimately to be caught in a freeze-up 60 miles downstream from its destination. Dog teams were called into service to transport piecemeal, these ice-bound supplies, over the ice, from the stranded boat to the construction site.

On the other hand, it was reported by the Dawson Daily News, at that time, that the *first* cargo of freight consigned to this railway arrived in Dawson City, on board the river steamer *Canadian* carrying 50 tons of rail and 150 tons of camp supplies and grading equipment. This tonnage was received in transfer from the White Pass Railway in Whitehorse. Except for the rail, this load was subsequently transferred, several days later, from the *Canadian* to the riverboat *Light* for the 1,700 mile trip down the Yukon River and then up the Tanana River to the construction site along the Chena. Why rail was left behind in Dawson City for an indefinite period is uncertain. Perhaps the late seasonal start in construction of the railway preempted any possibility for track-laying that autumn, or maybe the water levels on the rivers receded to such a degree, that full tonnage could not be carried. Never-the-less, although other vessels continued to load up in Whitehorse at this time for the long cruise to Chena, by August 31, 1904 all such movements would come to a halt. The fear of low water and imminent freeze-up preempted any further shipments of freight until the spring of 1905.

With the arrival of spring of 1905, the inevitable river breakup was to follow. Boats which had been locked in the icy grasp of winter, once again pressed on downriver to the newly developed mining camps. By May 16, 1905, the rush was on to get the rails, rolling stock and other supplies to the site of the railway construction along the Chena-Fairbanks route. One of these river steamers, the *Columbian*, which was forced to winter in *Hootalinqua*, broke loose from the ice pack and delivered its loaded railway iron and other freight to Dawson.

Hootalinqua was a staging area located in the Yukon Territory at the confluence of the Yukon and Teslin Rivers at a point just north of Whitehorse. Since there were over 800 tons of freight still sitting at this location and still more coming up from Whitehorse, the *Columbian* made an additional return trip to Hootalinqua for additional rail and other freight to be moved downriver to Dawson. Because of water level fluctuations in the river at this time of year, the prevailing depth of water at any one location would limit not only the size of the steamer, but also the amount of tonnage that it might carry. Thus, White Pass vessels coming downriver to Dawson from Whitehorse, were able to handle only two-thirds of their maximum tonnage. This was especially so in the region of the Fiftymile and Thirtymile rivers. However, beyond Hootalinqua full loads could be handled to Dawson. It was for this reason that Hootalinqua was made a staging point in the autumn of 1904, for the freight that was to move on to the Fairbanks region the following spring.

By the middle of May, 1905, river steamers began arriving at Chena with the materials for the new railway. These materials arrived from Dawson aboard steamers belonging to the Northern Commercial and the North American Transportation Companies. In this armada of steamers plying the Yukon River at that time, delivering rail to the Tanana Mines Railway, was the famous *Lavelle Young* of Barnette fame. Along with its 80 tons of rail, it carried 42 tons of *Beer*!

Paragraph 1 consolidates references: Senate Congressional Report - Joslin 1909;247; April 1912:79 and Dawson Daily News articles dated, May 16 & 20, 1905

Coming over the ice and snow again from the Yukon, but this time from Whitehorse, Falcon Joslin arrived in Chena on April 4, 1905 in company with some key people needed to build the railway. Members of this party included Charles Moriarty, Superintendent of Construction; Robert Taylor, Accountant; and Adam Tyson, General Storekeeper. Later came John Barnard, the Transit Man; L.L. James, Treasurer; and two time keepers. Quoting Martin Harrais, "This was the overhead force during the construction days — eight men." Arriving at a later date was James H. Rogers of *White Pass Railway* fame whose multi-faceted role was to appoint the various agents for the railway, and to develop appropriate and serviceable passenger and freight schedules. Following somewhat later was the first General Agent of the line, J.H. Scott. He arrived on the riverboat *Rock Island #1* somewhat before the completion of the new railway depot.

Paragraph 2 references an excerpt from *This Old House*

Charles Moriarty, the Superintendent of Construction, was singularly, one of the key players in this whole enterprise. His prior employment was that of Superintendent of the *White Pass Railway*. He became known as the "Snow King" because of his legendary exploits as he had battled, earlier in his career, the high snows found along the right-of-way of the Great Northern and White Pass Railways. His task on the Tanana Mines Railway was not only to fight off the severe yearly snows, but to construct and maintain this line during a period which would extend from the placement of the first tie until the time that he resigned in the spring of 1908. He was considered, by all that knew him, as "one of God's own gentlemen," and "probably was the most widely known and popular railway man in the north." He died in the spring of 1908, in Seattle, shortly after leaving the employ of the Tanana Mines Railway.

CHARLEY MORIARTY DIES IN SEATTLE

News reached this city this week of the death of Charles Moriarty in Seattle, probably the most widely-known and popular railroad man in the North. The sad news caused many expressions of sorrow to be uttered - as Moriarty was one of God's own gentleman, and God knows how to make a gentleman, too, when He wants to. His death resulted from an operation for stomach trouble, and, although usually a very strong and robust man, was unable to recuperate from the effects of the surgeon's knife.

Mr. Moriarty had the great misfortune to lose his wife in St. Joseph's hospital at Fairbanks on February 4th, last year. He, in company with three of his children, took the good woman's remains out over the ice via, Valdez to Roseburg, Ore. - their outside home. Shortly after their arrival in Oregon, the youngest, Eileen, aged 7 years, never a strong child, followed its mother to great beyond. It was rumored, that, a few months ago the older daughter, Maureen, had also died, but the report is unconfirmed. Mr. Moriarty was for several years superintendent of the White Pass railway, where he was nicknamed the "Snow King," and in the early spring of 1905, he journeyed over the ice from Whitehorse to Chena with Falcon Joslin, and was superintendent of the Tanana Valley Railway from the time the first tie was placed in position until he resigned last spring.

Paragraph 3 references Hot Springs Post, May 1, 1909

With the onset of spring 1905, Moriarty got his work crews together and set his work plan into motion.

Newly laid track. Note the unfinished ties and the absence of tie plates. Also, nowhere is there a locomotive to be seen on this construction site.
Photo courtesy of the Paul Solka Collection

Looking upstream at Chena, along the Tanana River. The engine house for locomotives of the Tanana Mines Railway/Tanana Valley Railroad are located just to the left of the docked riverboats.
Photo credit: University of Alaska Fairbanks, Alaska and Polar Regions Department Acc. #58-1026-2211N, in the archives, of the Bunnell Collection

Paragraph 5 consolidates references a letter from Falcon Joslin to Riggs, A.E.C., September 21, 1915, and Dawson Daily News, November 18, 1904

Two hundred men, *gandy dancers* (track layers), construction men and laborers from all works of life were recruited, at $7.50 a day. These wages were considered to be, at that time, the highest in the world and were forced upon Falcon Joslin by the eternal necessity of attracting and retaining men from the placer mines. Once workers made their "poke" it was back to the mining creeks with a new "outfit" on their backs.

Using no federal subsidy, the Tanana Mines Railway began its construction with a sharp resonance of steel on steel that echoed across the flats of the Chena River. Headquarters for the railway was established in Chena, a position that it held until 1915. Resting along the shores near the confluence of the Chena River with the Tanana River, the railway yards were located to the northeast, while the town of Chena and the river docks were to the west, paralleling the Tanana River. The railway terminal yards, which followed along the course of the Chena River, comprised 18.5 acres of land. Leaving these yards and running generally in a westward direction, 0.67 miles of track ran down Main Street into Chena town and then down to the river docks.

Having completed the track for the yards and dock in Chena, Moriarty and his men started construction, in a northeasterly direction, of the line that would ultimately go to Fairbanks and the mining creeks. Using locally cut Spruce timbers, ties for the track and bents and stringers for the bridges were made. Ties measured six inches by seven inches by six feet. Some of the ties were cut on all four sides, while others were smooth-cut only on two sides with the tree bark remaining on the other two. Forty pound rails — that is, forty pounds to a yard, the standard for the line — were held in place with two, five inch railway spikes per tie. These spikes, with a shank width of ½ inch, were much smaller than those used on the standard gauge lines. No tie plates were used.

From Chena, the line progressed in a northeasterly direction, almost in a perfect tangent, for five miles. It passed through low, swampy land and muskeg, subject to flooding that was further aggravated by the underlying *Permafrost*. Except for the line traversing the hills between Gilmore and Olnes, this same set of hazards prevailed along the rest of the railway.

At the time that track laying began, it was envisioned that few, or no problems would be encountered, as it traversed flat terrain. However, although the topsoil could be removed with little difficulty, the structure of the underlying surfaces at this latitude, presented far more civil engineering problems than were initially contemplated. The surface was covered by one to three feet of moss, under which there was primeval frost. This frost and ice could extend down into the ground for as much as 200 feet below the surface, or even farther. It could never be extricated from the ground. In some places, clear ice was covered only by a few inches of muck, vegetable mold and moss. When the moss was broken and stripped from the ground, the muck and ice began, at once, to melt. In spite of the lush vegetative

a Valley — appearing in places like a northern jungle — hidden just below the ı the valleys, was the ice, better known as *Permafrost*. This forms when the mean 's approximately 27 degrees (F). Fairbanks has a mean temperature close to 27 :, has these "islands" of frozen ground. Vertical ice lenses occur where the ground d with water in a climate where the temperature falls below 0 degrees (F) for a ime.

Alaska Science Nuggets, pgs. 70 & 74

ıf the ground was maintained by the moss which protected the permafrost from The thawing and melting caused by moss removal soon made a quagmire out of ıtain the integrity and grade of the railway surface, immense quantities of gravel ese locations to prop up the roadbed and track. With temperatures varying from : winter to plus 90 degrees in the summer, environmental havoc was played on ng process that occurred in the summer could be as much as two feet. As a r embankment of gravel on top of these permanently frozen areas was just like ss a piece of ice. This caused extensive undulation of the line and required ıd reballasting. As a consequence, the elevated embankment would vacillate : ground, to that of only a foot or so.

Paragraph 1 references a report entitled Railroad Engineering under sub-arctic conditions, written by Anton Anderson, February 14, 1964

Paragraph 1 & 2 references a letter from Falcon Joslin to Riggs, A.E.C., September 21, 1915

ıt the permafrost problem a new approach was devised for the construction of ɛaking or stripping the moss from the plotted route, ditches were cut, parallel e roadbed. The track was laid *directly on top of the moss*. Melting of the but instead of forming a quagmire, the run-off water drained into the ditches ıd track dry. However, this did cause considerable settling of the track as the cheese-like effect on the base. These areas were then ballasted with gravel solid foundation for the railway; whereas, the trains were initially riding on a jelly-like surface.

Having reached a point which was five miles from Chena, the railway split off in two directions; this location is almost due south from the present headquarters of the University of Alaska, Fairbanks farm complex. One line turned due east, which would take it to Fairbanks, while the other line turned to the west, which ultimately, took it to Fox, Gilmore and Chatanika. Both of these lines were interconnected to form a perpetual "Y" formation. Though the line to the east is often referred to as the "branch line to Fairbanks", this is not entirely correct. Early during the inception of the railway, Falcon Joslin recognized that although Chena may or may not have represented the best location for navigation, the ultimate destination for people and cargo was Fairbanks. In addition, there was fear of the possibility of another competing railway line being built between these two communities which would greatly compromise the Tanana Mines Railway. Therefore, it was elected to build the line from Chena to Fairbanks

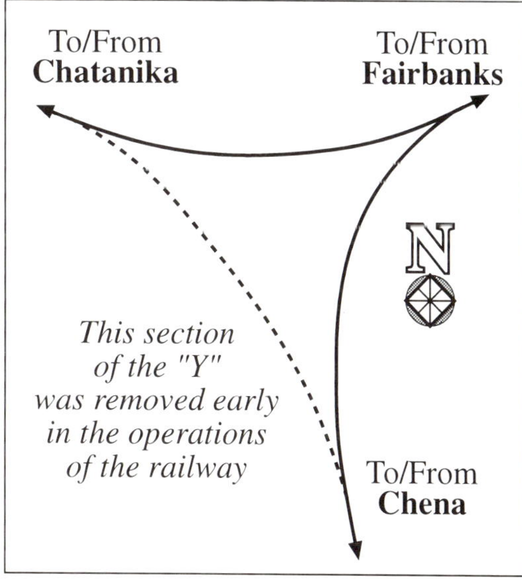

Author's Note:
The original plan for the construction of the Tanana Valley Railroad, as formulated by Charles Moriarty, Superintendent of Construction, designated the line from Chena to the Junction as the *mainline*; the line to Fairbanks from this point was to be considered a branch line. It was soon realized that potential competing lines and increasing tonnage being deposited at the docks in Fairbanks by the riverboats, required a change in the operating plans. The switches at the Junction taking the track from the Chena line to the Gilmore line were taken out. All trains out to the creeks from Chena were thus forced to go first to Fairbanks. The trains would then reverse themselves and return to the Junction finally on its way out to the creeks.

Paragraph 1 references an article from the Fairbanks Evening News, August 1, 1905

Excursion train crossing Noyes Slough shortly after leaving Fairbanks. Present day Alaska Railroad is just a few yards to the right of this long gone structure. Photo credit: University of Alaska Fairbanks, Alaska and Polar Regions Department Acc. #70-28-615N, in the archives, of the Wilson F. Erskine Collection

Paragraph 3 references an A.E.C. report, March 12, 1914 through December 31, 1915, by W.S. Johnson, General Manager

Paragraph 4 references an article from the Fairbanks Daily Times, August 17, 1907

as the first priority of the Tanana Mines Railway.

Further, most freight from Chena out to the mining creeks first went to Fairbanks and then backtracked to the Junction before going out to the mining sites. In time, the successor to theTanana Mines Railway, the Tanana Valley Railroad, was quite often affectionately referred to as "The Old Reliable Tanana Valley Railroad", while that portion of the railway that went from Chena to the Junction was referred to, rather derogatorily, as the "Dummey Line"! In time, the direct line at the "Y" from Chena to the creeks was removed.

Comprising an area of 40 acres, the Junction was referred to by a variety of names. Depending upon the traveler's destination, it was referred to as, Fairbanks Junction, Chena Junction or Ester Junction. No station building was present there.

From the Junction, the railway took a turn of approximately 90 degrees to the east and proceeded down a low, tree covered valley for two miles on a slightly elevated embankment which was devoid of any ballast. The railway then crossed Noyes Slough. This location was the source of continuous engineering headaches which was predicated, primarily, on haste of construction. The structure was a wooden trestle which was built quickly to meet Falcon Joslin's intention to get the railway in operation as soon as possible. Many compromises in construction were made which resulted in constant repair and replacement of timbers. Spring thaws caused the fracturing of large destructive islands of floating ice which easily tore out gaps in the bents of the bridge and made the bridge unsafe and totally useless.

As the railway approached its ultimate terminus at Garden Island, it crossed the Garden Island Slough. Approximately 350 feet prior to crossing this slough, the railroad built, in 1907, an additional track that was 2,500 feet long. It served the Noyes Lumber Mill which was located on the slough of the same name. Finally, the line made a 90 degree turn to the south over one of the sharpest curves a railroad could traverse. This curve still exists today on the Alaska Railroad as it enters its final approach to the station! The Tanana Mines Railway proceeded for an additional several hundred yards to the very edge of the north bank of the Chena River. This terminal complex of 19.802 acres was not actually in Fairbanks, which rested on the south bank of the river, 400 feet beyond the end of steel, but rather in *Garden Island*!

A warehouse, engine house and a two-story station measuring thirty feet by fifty feet were built here. The railway ran on the west side of the building. The east side was reserved for a garden to grow flowers and vegetables, a hall-mark of green-thumb, present-day Alaskans. The first floor of the depot was divided into a public waiting room, a ticket office and a baggage room. The second floor was partitioned off into offices and living rooms. The offices were not only for the railway brass, such as

Falcon Joslin, but also for other government agencies.

This depot was located on a site at North Cushman and Brandt Street, approximately where the present-day parking lot for the *Fairbanks Daily News-Miner* is located. It was actually built before the arrival of the railway, in 1904. The outside finish of this two story frame building was painted a rustic red color. The inside finish had double walls, one inch shiplap native spruce covered with cloth and paper, and was insulated with sawdust. Each story had ten foot, six inch ceilings, with a roof made of corrugated metal.

Wade Joslin, (a nephew of Falcon Joslin) who was the first President of the Tanana Mines Railway, was the first agent to occupy the apartment on the second floor. At a later date, his position was taken over by John Raap, who became the local railway agent and new occupant of the apartment. Raap continued on as agent until 1917, the last year that the Tanana Valley Railroad, earlier known as the Tanana Mines Railway, existed as a corporate entity.

South end of Fairbanks yards. Note the dual tracks. The two rails to the right belong to the Tanana Valley Railroad, while the two outside rails belong to the Alaska Railroad. The highway bridge to the left crosses the Chena River and is now in use at Nome. The International Hotel is still standing.
Photo courtesy of the Candy Waugaman Collection

Opposite the depot was a complex of business buildings parallel with the track, which even today bear, with some modification, the same general layout as at the time of the gold rush. Initially, this string of buildings consisted of Samson's Hardware, the Alaska Citizen's Office and Print Shop, the H. C. Davis Sash and Door factory, Miner's Home Saloon, Garden Island Grocery and the International Hotel, which is still standing and serving its patrons.

At the warehouse, which stood some distance from the depot, the track also ran along the west side of the building. On the east side was a parallel dirt road which facilitated the transfer of cargo and freight from one transportation mode to another.

Though the railway had to concede to the Chena River in its plans to enter Fairbanks proper, it did so after much effort, time and money was invested in the project. It actually built a pole trestle across the river only to have it destroyed by the flooding waters shortly after its arrival in Garden Island in 1905. The railway's intention was two fold: first, to tap the new mining strikes down the Richardson highway in the Sourdough District; and second, to continue down Cushman Street to the Tanana River, with the intention of developing another access to the river where plans were made to construct some new docks. Track was actually laid from Turner Street to 12th Avenue.

Dawson Daily News, July 2, 1905

However, by the time the Tanana Mines Railway had reached the north shore of the Chena River, a foot and wagon bridge had already been built across the river. This, the first Cushman Street bridge, was 305 feet long and had a forty foot draw bridge to allow the passage of riverboats. This draw bridge was also raised during breakup so as to reduce the damage to the bridge itself. It crossed the river pretty

Alaska Road Commission report, 1917

much at the location of the present bridge, that is, from the corner of First and Cushman Street over to Garden Island. This was the beginning of navigation, downstream, along the Chena River.

On July 4, 1905, Falcon Joslin planned to have a large celebration to commemorate the "completion" of the railway by driving a *Golden Railway Spike*, instead of the usual steel one, into the final segment of track. It was the consensus of the public that the railway, which was projected to go as far as Gilmore, would be completed by July 15, 1905. Joslin had recognized the importance of the Chena-Fairbanks run as being imperative for the immediate movement of passengers and freight — but, by the same token, he recognized the importance of getting the line into operation soon, if for no other reason than to instill a measure of public confidence in the venture.

The Golden Spike ceremony marking completion of the Chena-Fairbanks segment of the Tanana Mines Railway. The event occurred on July 17, 1905. The spike was driven into the tie by Isabelle Barnette, wife of E.T. Barnette, who stands to the right in the photograph, holding the spike in her right hand.
Photo credit: University of Alaska Fairbanks, Alaska and Polar Regions Department Acc. #76-3812N, in the archives, of the Terry Cole Collection

Paragraph 2 references *Flag Over the North*, pg. 297

As the momentous day arrived for the Golden Spike ceremony, an engorged Chena River, fed by spring rains and melting snow, came cascading into Fairbanks, spilling over its banks to cause much destruction and mayhem. A 250 foot river dock, just completed the day before, on July 3, by the Northern Commercial Company, was torn away from its banks and completely destroyed, as was the pole trestle constructed by the Tanana Mines Railway. The projected line down the Richardson to the Sourdough District was now severed and scrapped as a project.

Paragraph 3 references an article written by Judge James Wickersham and published in *Old Yukon*, pgs. 475, 1938

Once the water receded and structural damage to the railway was repaired, plans were rekindled to hold the Golden Spike ceremony. On Monday, July 17, 1905, at 2:30 P.M., in the midst of a great number of dignitaries, *two ceremonies took place*. Initially, the affair was opened with an introductory speech by Judge Carr in which he presented to Judge Wickersham the *first* spike ever driven into a tie of the Tanana Mines Railway. Then the moment of jubilation occurred — Mrs. Isabelle Barnette, the wife of E.T. Barnette, wielding a spike-driving maul, drove a forty ounce Golden Spike made from locally mined gold dust into the tie to formally "complete" the Tanana Mines Railway! These festivities took place at the "end-of-steel", in the Garden Island railway yards opposite to the recently washed-away bridge over the Chena River.

Fairbanks Evening News, August 7, 1905

Shortly after the Golden Spike was driven into the tie to commemorate the completion of the Chena-Fairbanks segment of the railway, it was pulled out — again! On Saturday, August 5, 1905, during a special banquet, it was presented to Mrs. Isabelle Barnette. On the shank of this railway spike was inscribed, *"Driven by Isabelle Barnette!"*

Following the spike-driving ceremonies, the management of the railway opened up the facilities to the public. Locomotive No. 1 and a recently acquired passenger car were hitched together with their "link and pin" couplers and made available for the people of the community to take short, get acquainted excursions. This locomotive was not only the first engine on the railway, but it was also destined to be the only surviving power unit from the Tanana Mines Railway to this very day — and, into the twenty first century! Locomotive No. 1 arrived in Chena just in time for the 4th of July, 1905 celebration, which was derailed because of the ensuing Fairbanks flood. Since the railway construction began in the late autumn of 1904, and then, once again, re-started in late spring of 1905, it must assumed, if one looks at the arrival date of locomotive No. 1, that no motive power was available during the construction days to haul men, rails and ties. In all probability, these supplies must have been carried on flatcars, which had earlier arrived in Chena from Dawson City — to be progressively pulled along the recently constructed track by beasts of burden such as horses and mules.

Fairbanks Evening News, August 24, 1905

Locomotive No. 1 was a small saddle tank engine designated as a 0-4-0. This meant that it had four large wheels to drive the unit, but with no leading or trailing trucks which was common on larger locomotives. It was built by H.K. Porter Company of Pittsburgh, Pennsylvania, during March of 1899. Although it was referred to as No. 1 by the Tanana Mines Railway (Alaska Railroad), its original serial number was 1972. It was purchased new by the North American Transportation and Trading Company for its coal mining operations at Cliff Creek. This mining enterprise was located along the Yukon River, 59 miles downstream from Dawson City near the Fortymile country. The locomotive came north, presumably by sea, to St. Michael where it was put aboard one of the riverboats belonging to the North American Transportation and Trading Company, during the summer of 1899, for delivery to the one and three-quarter mile long railway at Cliff Creek. This riverboat company had its headquarters at Hamilton, at the mouth of the Yukon River, and operated seven steamers between St. Michael and Dawson City. It was owned by the infamous John Jerome Healy, who through correspondence with E.T. Barnette, had related his plans to cover Alaska with a system of railways, striking at the same time, Barnette's imagination to make his trip up the Yukon and Chena Rivers.

Locomotive No. 1 the first engine on the Tanana Mines Railway. The location is uncertain, but appears to be at Chena Junction. Note the link and pin coupler.
Photo courtesy of the Candy Waugaman Collection

Mining Railways of the Klondike, pgs. 6 & 7

Between 1899 and 1903, this locomotive worked the mines at Cliff Creek by hauling the coal down to the river's edge for ultimate transport, by riverboat, to the Dawson Electric Light and Power Company — a Falcon Joslin venture! Once these mines became no longer productive, the locomotive was sold to the Coal Creek Coal Company which was located five miles upstream from Cliff Creek. This was another operation in which Falcon Joslin played an instrumental role in its development. On August 4, 1903, the steamer *S.S. Bailey* transferred this locomotive from Cliff Creek to Coal Creek. Locomotive No. 1 worked this 12 mile long, coal hauling railway, from August, 1903 to June, 1905.

Mining Railways of the Klondike, pgs. 17 & 21

On May 31, 1905, the steamer *Louise* departed from Dawson City with 350 tons of cargo consisting primarily of rails for track construction along the Tanana Mines Railway, and also railway cars for the same line. Early in June this steamer stopped at Coal Creek and picked up the diminutive locomotive

No. 1 for use on the Tanana Mines Railway. Originally, the engine was to be picked up earlier in the season by the *Cudahy*, but large ice flows in the Yukon River prevented this transfer. All this freight which had now swelled to 600 tons was placed aboard two barges which were lashed to the *Louise*. These barges were taken as far as Ft. Gibbons (Tanana Village) where they were picked up by another river steamer for trans-shipment to Chena. The *Louise*, in the meantime, went on to St. Michael. Locomotive No. 1 was unloaded from the barge in Chena to start service on the Tanana Mines Railway, the Tanana Valley Railroad and on the Alaska Railroad until 1923.

Dawson Daily News, June 1, 5 & 17, 1905

~

The first engineer on locomotive No. 1, John Wigger, also known as *"Moose John,"* ran his "train" from Chena to Fairbanks and returned every hour. While many of the sourdoughs took advantage of this offer to ride the train, others found themselves in an incomprehensible dilemma — that is, their desire to ride the train was in conflict with a fear of the "Iron Horse!" Some of these glad-faced sourdoughs hadn't had a ride on a railroad car for as long as eight years, with the result that many returned from their excursion complaining, real or imagined, of liver and kidney complaints — which in all probability were self-induced celebration troubles of their own making. Others were too frightened to board the "train"! Blindfolds were applied to their eyes and, very gingerly, they were backed into the waiting passenger car for the six mile ride down the pike.

Paragraph 1 consolidates references the Chena Times, July 18, 1905, and *E.T. Barnette*, pg. 100

On the following day, Tuesday, July 18, 1905, the Tanana Mines Railway hauled its initial freight from Chena out to the Four Below Commercial Company. While some of it was deposited along the line at various points, the remaining freight was delivered to Fairbanks. The advent of this railway service was to have a negative effect on the "Mosquito Fleet" which plied the Chena River. This flotilla of light river-craft came into existence before the arrival of the railway as a contingency service between the docks at Chena and Fairbanks, should the water level of the Chena River be too low to support steamer travel along this route. This service became doomed when the railway initiated a schedule of four trains-a-day between these two points, maintaining a "mixed service" whereby both passenger and freight cars were utilized on each train. The stretch of track between Chena and Fairbanks became the domain of locomotive No. 1 for it was well suited for the level conditions that existed between these two communities. In addition, its other role was to shunt the cars in both Chena and Fairbanks, never venturing off elsewhere on the line.

Paragraph 2 references Chena Times, July 18, 1905

TANANA MINES RAILROAD CO.
Schedule of Trains

Until further notice trains will run as follows:

No. 1 - Leaves Chena Daily	8:00 a.m.
Arrives Fairbanks	8:40 a.m.
No. 2 - Leaves Fairbanks	11:00 a.m.
Arrives Chena	11:40 a.m.
No. 3 - Leaves Chena	3:00 p.m.
Arrives Fairbanks	3:40 p.m.
No. 4 - Leaves Fairbanks	5:00 p.m.
Arrives Chena	5:40 p.m.

The company reserves the right to change this schedule at any time without notice.

C. Moriarity, Supt. **J.H. Scott Gen. Agt.**
FALCON JOSLIN, V.P. & G. Mgr.

1905 Train Schedule printed in Fairbanks Evening News, August 4, 1905

Tanana Mines Railway mixed train at the Chena Junction, which is today the approximate location of the agricultural farm at the University of Alaska, Fairbanks. The train is coming up from Chena and is on one leg of the "Y" which leads to Fairbanks. Notice the short 19 foot, 4 wheel boxcar on the end. The track in the foreground leads to the mainline to Sheep Station (Happy). This line was to become the Alaska Railroad of today.
Photo Credit: University of Alaska Fairbanks, Alaska and Polar Regions Department Acc. #79-41-45, in the archives, of the Falcon Joslin Photograph Collection

Locomotive No. 1 at the Chena Station making ready its consist for the run to Fairbanks. Notice the first car is a heated boxcar, while the second car is the "shorty" nineteen foot version.
Photo credit: University of Alaska Fairbanks, Alaska and Polar Regions Department Acc. #79-41-46N, in the archives, of the Falcon Joslin Photograph Collection

T. M. RY
50
Construction
M.RR Near the termines on Pedro Creek

CHAPTER 4

Chena Junction To Gilmore

Tanana Mines Railway Locomotive No. 50, an American 4-4-0, delivering ties to a construction site near the terminus of the line at Pedro Creek (Gilmore), milepost 20 on the original line. Locomotive No. 50 was the second engine acquired by the railway. Both flatcars were obtained from the Klondike Mines Railway in the Yukon, and still retain their original lettering in this photo.
Photo courtesy of the Candy Waugaman Collection.

Railroad yards in Fox, Alaska, 1911. Note the "Stub" switch off to the left. Switches of this type would eventually be outlawed.
Photo courtesy of the Olga Steger Collection

As important as the railway was to the early and subsequent development of Fairbanks, road competition was present right from the onset of this service. The principal competitor, at this time, was a transportation firm known as the *Clough-Kinghorn Company*. In addition to daily wagon service out to the mining regions, they had three pack trains which they used for small shipments, or when extra rapid delivery was desired. Although they advertised that no order was too large for their equipment or too small for their patience, it was their inability to deliver *tonnage* freight at a reasonable cost, that made the railway a vital necessity.

Paragraph 1 references Fairbanks Evening News, August 3, 1905

Paragraph 2 references Fairbanks Evening News, August 7, 1905

It was because of this actual competition from road traffic and of perceived competition from a parallel railroad from Chena to Fairbanks that Falcon Joslin gave such priority to the newly constructed line between these two points. By August 7, 1905, the railway had laid only six miles of track from the Junction out to its destination — Gilmore! Perhaps the slower construction pace for this portion of the line could have been predicated to the fact that railway supplies were still arriving on a piecemeal basis and an itinerant schedule.

Locomotive No. 50, an American 4-4-0, splits a large mass of lumber in the Tanana Mill yard, in Fairbanks. Photo credit: University of Alaska Fairbanks, Alaska and Polar Regions Department Acc. #70-58-373, in the archives, of the Ralph Mackay Photograph Collection

On June 19, 1905, the steamer, *Columbian*, arrived in Dawson City after coming down the Yukon River from Skagway. On board was locomotive No. 50 and a passenger car, both destined for the Tanana Mines Railway. The locomotive, an American type 4-4-0, made by the Baldwin Locomotive Company had seen service on the White Pass and Yukon Railway. These two pieces of equipment, plus several flatcars and rail, left Dawson for Chena on June 19, 1905. Other boats followed with supplies. One of these boats was the *Cudahy,* which was flying the flag of the North American Transportation and Trading Company. On August 2, 1905, it arrived in Fairbanks with 200 tons of railway iron after having made a long trip from St. Michael. Following the *Cudahy* by about a week was the riverboat *Power* which arrived on the Fairbanks scene on August 9, 1905, with another 200 tons of railway material.

Paragraph 3 combines references from Dawson Daily News, June 17, 1905 and Fairbanks Evening News, August 1 & 7, 1905

At the same time that all this "new" equipment and rail was arriving in Chena and Fairbanks, route changes were already being contemplated. Although the lines immediate destination was *Gilmore*, gold strikes at *Cleary* and *Fairbanks Creeks* gave the management of the railway second thoughts. Even while laying track out towards Gilmore, the Tanana Mines Railway began a preliminary survey over the Cleary ridge for a line which would carry it into the mining camps in the Fairbanks and Cleary Creeks. All this was going on at the same time the "gandy dancers" were laying track as fast as they could out to Fox and Gilmore. Even as late as the following August of 1906, this plan was still being contemplated by Civil Engineers in the employ of the railway, who were surveying a route from *Cleary Summit* to the end of the line at Gilmore. Although it was reported that there was nothing in the way which would make the construction of this line difficult, it was ultimately decided to take a more circuitous route over to the other side of the hills. While the construction of the line out to Gilmore was still taking place, work was started on a telegraph line which paralleled the railway track from Chena to

Fairbanks Evening News, August 7, 1905

Fairbanks Daily Times, August 29, 1906

Fairbanks and from there continued out to the creeks. The railway, however, controlled train movements by a telephone system rather than by the conventional telegraph system.

Fairbanks Evening News, August 7, 1905

Fairbanks Evening News, August 1, 1905

It seems that the delay in laying down the rails to the mining creeks was due, in actual fact, to their *non-availability*. Up to the point of depletion of the rails, Superintendent "Snow King" Moriarty had his crew putting in track at the rate of a mile a day. Never-the-less, cutting and grading of the roadbed, such as it was, continued. From the Chena Junction, milepost 5, at an elevation of 526 feet, the line proceeded northward along St. Patrick's Creek. At milepost 6, just opposite the present day Smith Lake, the railway installed a tangent track and a 240 foot spur which became known as *Ester Siding*. Located here was an outside braced station — but without any agent. To obtain a reduced rate on their tickets, travelers to this location, and all other non-agent stops, were encouraged to buy round-trip tickets from either the Chena or Fairbanks agent, 15 days in advance.

Ester City, a newly discovered mining region, was 3.5 miles from Ester Siding and had no direct rail connection to the mainline. It had two hotels, the Golden Eagle and the Occidental, in addition to several roadhouses. Many other businesses, such as merchandise stores, restaurants and the usual big compliment of saloons, were scattered throughout the local mining area. Surrounding this enclave of businesses were mining operations located at 3 below and 5 below, which represented their position in respect to the original discovery Claim. That is to say, a location of 3 below would represent a mining operation which was located 3 claims below the original Discovery Claim.

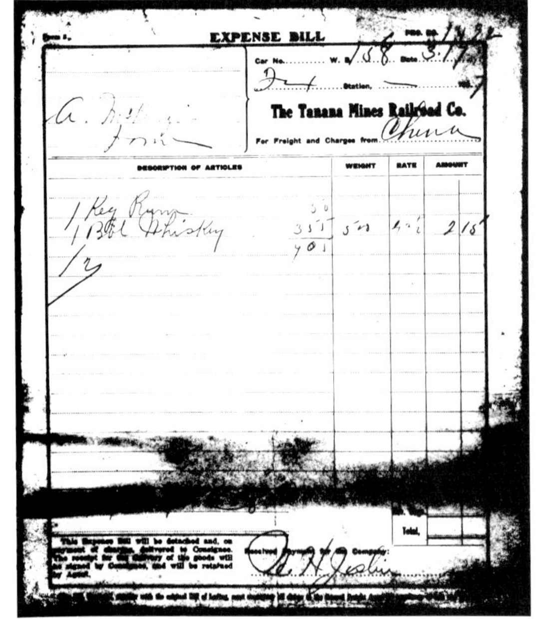

EXPENSE BILL

Car No. .... W. B. 158 Date 3/1/...

Station, ...

The Tanana Mines Railroad Co.

For Freight and Charges from Chena

DESCRIPTION OF ARTICLES	WEIGHT	RATE	AMOUNT
1 Keg Rum 1 Bbl Whiskey			2 15

Notice the contents of this freight manifest. Review of other manifests indicate that, along with gold dust, this was a frequent freight item.
Photo courtesy of the Dick Farris Collection

All the surrounding camps were connected with Ester City, Fairbanks and all the rest of the Tanana mining region by telephone. Subscribers were admonished, however, not to listen in on other people's conversations — and to avoid using the telephone when it was thundering or lightening!

Since the intrusion into the wilderness by these miners was virtually overnight, the balance between man and the wilderness was, at best, very shaky. A miner from Ester Creek, named Mills, was out hunting one day for small game. All of a sudden, from behind the brush, a *big black bear* came charging at him. According to his hallucination, he claimed that the bear was fifteen feet long and seven feet high. Mills turned in his tracks and literally ran up the nearest tree. After looking things over, the bear casually sat down by this tree, looking very satisfied at *His* hunt — while Mills, on the other hand, was deep in prayer perched up on a precarious tree. Suddenly the bear arose from his sitting posture, grunted as he looked up at his potential victim, and slowly walked away. With great hesitancy, Mills took a full 15 minutes to climb down from the tree, after making it up there in nothing flat. Upon reaching the ground, Mills, it was claimed, broke all the world's records for the three mile run back to Ester City. He claimed that he covered the distance "in three seconds less than nothing"!

Fairbanks Daily Times, August 22, 1906

Passing through woods heavy with Spruce, Aspen and Birch trees, the railway proceeded to the east of diminutive Lake Killarney, reaching Happy, milepost 7.3 at an elevation of 672 feet. From this point, the line turned 90 degrees to the east and proceeded up the Goldstream Valley. Happy became very important as a railway junction, although fleetingly, later in 1919 as the point of departure for the narrow gauge line that was to be built to North Nenana. This railroad "town" was located on the north bank of the Tanana River across from the Indian village of Nenana. The 47 mile railroad was built by the Alaskan Engineering Commission as part of the grand scheme to connect Fairbanks to Seward by rail. By 1923, this line would be converted to the standard gauge of the Alaska Railroad whereby Happy, once again, would become an important junction; but this time as the point where the transition of track gauges took place.

Passing through the Goldstream Valley, the railway traversed an area of discontinuous permafrost — which meant that areas of the sub-surface were frozen year round. It is the land of the "little sticks" or Taiga, where islands of stunted Black Spruce dot the terrain. This frozen ground, or permafrost, occurs in the Arctic and Sub-Arctic when the mean annual temperature is 27 degrees F (-3 degrees C). As a result, excess water such as rain or melted snow, does not percolate into the ground. It either runs off the surface of the ground with a strong current into nearby rivers, or stagnates into pools of water forming a maze of ponds separated by soggy islands of muskeg.

Alaska Science Nuggets, pg. 70

Because of this expanse of ground which was saturated with water, the railway was faced each spring with an ordeal of riding over undulating track caused by extensive frost heaves. In some places along the line the track was submerged under water, under muck or displaced to one side or the other of the intended route. As a result, there was never any certainty that a train would arrive as scheduled, or at all, at its destination. It was precisely because of these circumstances that it became a virtual routine for the conductor, such as Jimmy Rodebaugh, or the brakeman, such as Tommy Hoover, to climb out onto the cow catcher of the locomotive and guide the train along. Holding a pike-pole in one hand and holding on to the locomotive with the other hand, he would extend the pole beyond the locomotive in search for the track as the train crawled along ever so slowly. This searching movement parallels that of a blind person in the use of his cane. On other occasions, the "pole man" would actually get off the locomotive, in hip boots, to wade through the water in search of the track. The ultimate scenario would occur later when the spring thaws suddenly brought down so much turbulent water on the track that it was totally and permanently displaced — and had to be reconstructed.

Tanana Valley Railroad mixed train passing through the flooded Chatanika Flats. Lead flatcar in the front of the train carries a "pole man" whose job it is to warn the engineer of any track damage caused by high water levels.
Photo courtesy of the Candy Waugaman Collection

Numerous wooden culverts were crossed as the line trekked through the Taiga. About a mile east of

Railroad yards in the modest town of Fox, Alaska, in 1911. Notice the "stub" switch off to the left. Switches of this type would eventually be outlawed.
Photo courtesy of the Olga Steger Collection

Happy, a forty-foot long wooden trestle was crossed passing over a small brook. Since mining claims seemed to be scattered all over the valley, the train would stop anywhere along the line to unload designated freight directly on the bare ground or at best, on a wooden platform. Some of these locations just off the railway were Carlson, Big Eldorado and Engineer — none of which had a station. Continuing on up the valley, the railway crossed what is now known as Ballaine Road. Passing over four small wooden culverts of various sizes, the railway crossed Goldstream Creek at milepost 12.5, traversing a forty-foot long wooden, through-truss bridge to reach the north bank of the creek.

Shortly after this crossing, *McNeers* station, a log structure, was passed. This station served the Big Eldorado mining district which was just north of this location. It was named after Arthur McNeer, a contractor, teamster and mining man. Since it was the distribution point for considerable freight into the surrounding hills, it had a substantial siding of 380 feet.

Paragraph 1 references Fairbanks Daily News-Miner, May 26, 1913

Paragraph 2 references Fairbanks Evening News, August 24, 1905

By August 24, 1905, the railway reported that it was about three miles from Fox, milepost 18, and about five miles from Gilmore, milepost 20, its ultimate destination. The company reported that it had plenty of material on hand to finish laying the rails, and that construction of the line would be completed by September 1, if, as Falcon Joslin stated, "No floods, earthquakes or other unnatural disturbances intervene."

Train "meet" in Fox, milepost 18, Tanana Valley Railroad.
Photo courtesy of the Candy Waugaman Collection

The line proceeded on through the muskeg of the Goldstream Valley to finally go into a slight climb. About 1.5 miles from Fox, the line began a climb out of the valley by skirting the shoulder of the high hill to enter the "city". Fox was, and still is, located opposite 14 below Discovery Goldstream, which at that time, was 12 miles by road from Fairbanks. The Tanana Mines Railway maintained a station here which had to be, eventually, enlarged to accommodate all the freight business. It also had an engine house in the yards where it based a helper locomotive to assist the train up the 2¼ percent grade to Gilmore, and eventually beyond. Located a short distance from the station was a water tower to replenish the steam-making abilities of the locomotives before they went up the "hill". The community acted as a distribution point for the surrounding mines and enclaves, and required the longest siding on the railway — 520 feet! In addition to the railway facilities, the "city" boasted of three hotels and three saloons along with four merchandise stores, a post office, a meat market and a public bath, all for a community of only 100 souls. Further, Fox had a laundry where they would wash, and deliver free, your garments anywhere in the immediate mining district. Lastly, it was linked to the general telephone system where you gave two cranks to call the attention of the operator to place your call, and one crank to indicate that your call had been completed. If one had to wait to get their call through, they could "rubber" (listen to other people's calls) to pass the time away!

Tanana Telephone Directory - 1907

After approximately five months of erratic construction in 1905, the railway was closing in on its final destination — *Gilmore*. Between Fox and Gilmore the pike assumed a posture more like that of a mountain climbing line. Leaving Fox, it crossed Fox Creek near the point where Fox and Pedro Creeks join to form the Goldstream Creek and proceeded up along Pedro Creek for approximately 1 to 1.5 miles. In 1907, a 180 degree massive wooden trestle was built at this location. It was then referred to as the "Loop." This "Loop" was built as means to approach the other side of the valley, gaining altitude in a confined region, to skirt the formidable hills to Chatanika.

Alaskan Engineering Commission Report, 1914-1915

However, in 1905, when this first section of the railway was built, the track would go on for another mile, while at the same time climbing a very steep grade to reach Gilmore, milepost 20. Ultimately, the last few hundred feet of this grade had to be leveled so as to expedite the transfer of merchandise from wagons to the train. Where originally only two wagons could be loaded at once, now it was made suitable to handle four wagons at the same time.

Fairbanks Evening News, July 17, 1906

Gilmore, which was named after Felix Pedro's partner, Tom Gilmore, was like all the mining towns at this time — it ebbed and flowed consistent with the fortunes of the miners. To service 450 people in 1907, it had a post office, three roadhouses, two of the ubiquitous saloons and two warehouses. Also, like all station stops along the railway, Gilmore had a stage line connection that serviced the Cleary and Fairbanks Creek mining regions. A 300 foot siding was located here to hold the increasing amount of freight which was beginning to arrive.

Paragraph 3 references Joslin 1909: 247 U.S. Senate Congressional Report, April, 1912: 7-9

With the arrival of the Tanana Mines Railway at Gilmore in September, 1905, the line, as originally visioned, was now completed. Twenty-two and seven-tenths miles were constructed at a cost of $20,000 a mile. No sooner had it been completed than plans were already in the mill for its "demise"! By 1907, it would become history, only to rise again, at the same time as the ***Tanana Valley Railroad***!

Locomotive No. 50, an American 4-4-0, prepares to depart Gilmore running backwards. This is the end of the track for the Tanana Mines Railway.
Photo credit: University of Alaska Fairbanks, Alaska and Polar Regions Department Acc. #79-41-47, in the archives, of the Falcon Joslin Photograph Collection

CHAPTER 5

THE TANANA VALLEY RAILROAD

A scheduled photo-stop for an excursion train on Trestle No. 6 at milepost 25.1. Trestle No. 6 was 502' long and 40' high at the deepest point.
Photo courtesy of the Candy Waugaman Collection

With an increase in mining activities now becoming a reality, especially out in the region of the lower Cleary, Fox and Chatanika Flats, Falcon Joslin felt it to be imperative to further extend the Tanana Mines Railway to these mining creeks. Unable to obtain any additional money from its original financier, *Close Brothers and Company* of London, England, the same investors in the White Pass and Yukon Railway, Joslin left on September 9, 1906, for Seattle and New York City. After extensive discussions with some new financiers, he was able to obtain financial help from the *Knickerbockers Trust Company*. In this endeavor, he raised $782,820 from both American and European interests. Thus, he was able to pay off the debt to Close Brothers and Company and start, with his remaining assets, the development of the extension to Chatanika.

Tanana Valley Railroad Annual Reports, 1908 & 1911

With all English interests, better known as the White Pass interests, paid off, Falcon Joslin wired his Superintendent, Charles Moriarty, on Monday, December 3, 1906, to make arrangements for the delivery of large amounts of railroad supplies, including ties and bridge timbers which were to be cut locally. Convinced of his own ability to raise the necessary money for this venture, he plotted, as early as the autumn of 1906, the permanent location of the new right-of-way.

Fairbanks Evening News, December 3, 1906

In actual fact, it was as early as 1905 that the Tanana Mines Railway began a preliminary survey over the Cleary Ridge for a line that would carry it to Cleary and Fairbanks Creeks. This survey was to continue into the autumn of 1906, as Civil Engineers were working from Cleary Summit to the end of the line at Gilmore. During this interim, however, mines came into existence over in the Dome, Vault and Olnes regions which changed the originally planned 12 mile extension over Cleary Summit into a 20 mile extension around the hill into Chatanika.

Paragraph 3 consolidates references Fairbanks Evening News, August 7, 1905 and Fairbanks Evening Times, September 7, 1906

On January 1, 1907, the property of the Tanana Mines Railway, originally chartered in the State of Washington, was transferred to its new successor, the *Tanana Valley* ***Railroad***, chartered in the State of Maine. The new entity took over all the physical properties of the Tanana Mines Railway, including its financial assets and liabilities.

Flushed with this bit of financial success, Falcon Joslin next approached the Congress of the United States for relief from the tax of $100 per operating mile of railroad. On March 2, 1907, the Congress replied favorably to this request and thus suspended this tax, as it applied to the Tanana Valley Railroad, for the next five years. However, what the government gave with one hand, it took away with the other hand. Already by the time that the railway reached Gilmore, it was beginning to see some hard times due to the dwindling supply of wood — a primary source of energy for the railway and for the whole society of the North. In an attempt to alleviate this impending crisis, management was looking into the prospect of tapping *coal* deposits which were relatively close to Fairbanks. However, by Executive Order of the President of the United States, the staking of coal claims was curtailed on

Tanana Valley Railroad Annual Reports, 1908 & 1911

November 12, 1906. Thus, the door was closed to a source of energy that could have supported mining and other developmental operations for many more years to come.

Al George, Land Management Office, May 8, 1981

To paraphrase Falcon Joslin, "If you want to get something done, then do it yourself," — and so he did! In order to get this railway built as soon as possible, he elected not to seek government aid or subsidy. As a result, he built the line with no grading of the roadbed and with little or no ballast to hold the track in place. Accordingly, the track became grossly distorted with many dips and rises superimposed by many "right angle turns." Fortunately, in most cases, the track gauge remained intact! Joslin's intention was to get the trains running as soon as possible, which would not only provide a service to the miners, but would also give him an early return on his investment. He paid for these indiscretions in that the resulting condition of the roadbed was so poor that the trains were forced to run very slowly. So slo-o-w, as a matter of fact, that it took an indigent ***Hobo*** to prove this point to the railway brass.

Fairbanks Daily Times, September 7, 1906

Charles Moriarty, the distinguished Superintendent of the Tanana Mines Railway, was riding on board one day on one of the Gilmore-Fairbanks runs. The crew was ever so vigilant in the performance of their duties, showing a great flare in displaying their diligence in the protection of the railway's interest, when a "bo" (Hobo) was discovered trying to beat his way onto a freight car for a ride from Gilmore down to town. Twice the train crew put him off. Finally, nothing more was seen of him until the train pulled into the Fairbanks depot. The Superintendent, quite satisfied with the crew's performance, was greatly astonished when he stepped from the platform of the coach to find the hobo nonchalantly smoking his pipe. His curiosity being aroused, Moriarty walked up to the hobo and inquired how was it possible that he was already in town. "How in earth did you get down here? I thought that you were put off the train at Gilmore", he said. "Say mister," replied the hobo, taking a deep draw on his pipe, "I don't want to knock your damned old train, but *I walked down*!"

Paragraph 2 references Fairbanks Evening News, July 31, 1906

With uneven track, marked by the absence of ballast, accidents occurred frequently. In many cases, trains would just ride over the rails and come to rest on their side. Fortunately, injuries were minimal.
Photo courtesy of the Colin MacDonald Collection

With an increasing demand placed on the transportation services of the railway in the face of limited equipment, plans were made to increase not only the number of train movements, but also to press for additional rolling stock. During the year of 1906, therefore, large expenditures were made to upgrade all temporary structures into permanent standards. In line with this urgency to update the railway, two shipments of much needed freight arrived during the early part of September, 1906.

The first shipment to arrive came on the steamer, *Seattle No. 3*. It carried six new railway cars for the line and a "new" locomotive — more powerful than those which were being used on the railway at that time. It was a heavier engine, built by the Brooks Locomotive Company, Mogul 2-6-0, locomotive No. 51. This unit was destined for the Gilmore section of track, for it was here that the railway grades were the steepest of any place on the line. The new locomotive was to be used for snow-plowing the right-of-

Fairbanks Evening News, September 8, 1906

way, and as a helper locomotive to assist heavy freight trains over the summit when the second phase of railroad construction was completed.

Arriving on the second steamer, shortly thereafter, was a big shipment of steel rails. Since the second phase of track construction out to the creeks and Chatanika was still on the drawing boards, this rail was destined to be used only for the maintenance of track and the construction of new switches and sidings.

While all this upgrading of the physical structure of the railway was taking place, an unprecedented bonanza of freight landed in Chena during that summer. Recently received flatcars were being rebuilt into boxcars so as to protect both the railway employees and the freight from the harsh and unbearable winter weather. In some cases, charcoal heating units were installed in the boxcars to preserve the perishables during their transport out to the mining creeks and camps. These unique cars were referred to as "hot-cars!" This great deluge of freight was due not only to the increased needs of the miners and the community at large, but more specifically, to the fact that river steamers were unable, because of low water, to travel up to Fairbanks. During the first week of October, 1906, only one steamer, the *Koyukuk*, was able to pass through the shallow Chena River to dock in Fairbanks.

Heavy smoke pours out of the stack of locomotive No.51 as it fights the heavy snow and frigid cold with its short train. The exact location is unkown, but is suggestive of the Goldstream Flats.
Photo courtesy of the Candy Waugaman Collection

Paragraph 2 references Fairbanks Evening News, September 8, 1906

Paragraph 3 references Fairbanks Evening News, October 9, 1906

The usual schedule of six trains a day was now increased to eight a day, with each locomotive pulling a capacity load. Further, the number of railway personnel was increased to double its original size. The relatively "big" locomotive, the Mogul 2-6-0, was meeting all the assumed expectations by being able to handle all these greater loads. It did, however, present some problems; its 48 inch driving wheels had some difficulty in rounding the sharp curves of the railway. Ed Ross, a machinist for the railway, resolved the problem by placing the wheels on a lathe and turned them down a half an inch. Not only did this procedure prove most satisfactory, but it also represented one of the biggest pieces of mechanical work ever carried out in this North Country. Trains were usually of a mixed consist — that is, in addition to the variety of freight cars that it hauled, it had an added passenger car attached to it, usually on the tail end.

Judging from the number of people all over the locomotive and cars, this must have been a scheduled photo-stop for rail fans and passengers. Locomotive No. 52 is a Mogul 2-6-0 and is pulling what must be considered an excursion train over the enormous trestle No. 6, at milepost 25.1.
Photo courtesy of the Candy Waugaman Collection

While the railway improvised extensively to dispatch the burgeoning cargo that was filling up the dock, it also found the means to accommodate the special needs of "tour" groups of passengers. Although fully accustomed to groups of sports fans and party people, it carried out on one occasion a rather unique charter. It seems that there was a land dispute between two miners, Klonas and Kelly, up on Dome Creek. Unable to fully comprehend the topography in this issue, the jury asked to be taken up to the site for a look-see evaluation. On Thursday, August 16, 1906, at 8 P.M., a special train carrying the jurors *(much like the political junkets of today)*, left Fairbanks for the closest railhead to Dome Creek. The rest of the trip was made by stage whereby the jury was most astutely able to carry out its public duties. In the meantime, a simmering train waited patiently for the return of its passengers, knowing too, that it had also met its public duty.

Fairbanks Evening News,August 12, 1906

Business had become so brisk that several newly constructed railway facilities along the line had to be altered relatively soon after their construction. Fox depot had to be enlarged to twice its size, for more and more freight was being delivered to this destination point. Now, whole carloads of freight, instead of the usual less than carload potpourri, were being switched off the mainline onto a siding. The freight was then transloaded to horse-drawn wagons and primitive trucks for distribution out to the mines along the Goldstream and Dome Creeks. In actual fact, it can be said that Fox had two depots during the course of its history.

Elsewhere along the isolated sections of the line, platforms were being built where stations were not warranted, but where freight volume was significant enough that it was not desirable just to leave it on the ground. Even in Fairbanks, the loading platform was extended 70 feet so as to accommodate increasing passenger loads.

Fairbanks Evening News, July 17, 1906

Typical mining enclave which was found throughout the gold bearing region.
Photo courtesy of the Nicholas Deely Collection

Some of the locals in Fox being held up by a passing dog. The railroad station is just off to the right of the picture. Note the railroad loading platform with the sled resting on top of it.
Photo courtesy of the Olga Steger Collection

BACKING UP FOR A RUN ON

CHAPTER 6

THE CONSTRUCTION OF THE TANANA VALLEY RAILROAD

Climbing a 2 ¼ percent grade toward Summit and Ridgetop, Mogul No. 52 needs the assistance of another locomotive to get its train through the heavy snows on the hilltops between Fox and Olnes.
Photo courtesy of the Candy Waugaman Collection

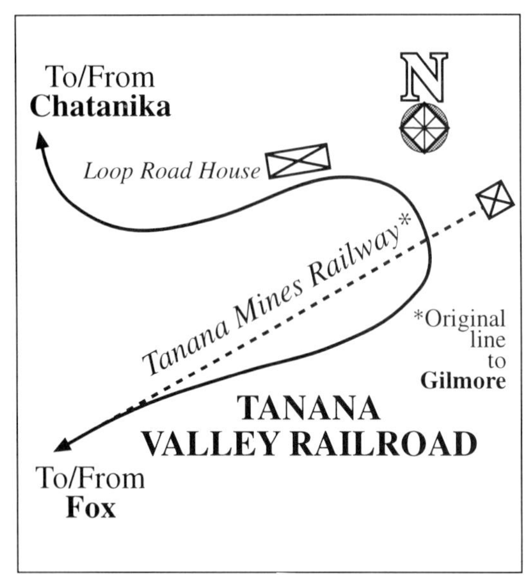

Paragraph 1 references *Alaska - Yukon Magazine*, December, 1907, pgs. 289 & 290

Paragraph 2 references Fairbanks Daily News-Miner, January 27, 1910

The second Loop Roadhouse at the north end of a curved trestle which the train used to reverse its direction as it climbed out of the valley below. Note the charred end of the building.
Photo credit: University of Alaska Fairbanks, Alaska and Polar Regions Department Acc. #76-92-83N, in the archives, of the Margaret Lentz Photograph Collection

On May 15, 1907, the second phase of the construction of the Tanana Valley Railroad began. This section of the railroad, which was originally planned to be a 12 mile extension from Gilmore, was to go up and over Cleary Summit to the mines present in the lower valley. While it was not certain as to whether or not the ruling grade was prohibitive to the summit, the fact that other mines were being developed along the alternative route may have been the impetus to route the railroad around the hills. Ridgetop, Olnes and the new town of Chatanika, which came into its own after the destruction of Cleary by fire, became established along this new line as mining and distribution centers. Costing $20,000 a mile to construct, this portion of the pike included ten wooden trestles, some of which were 600 feet long, and 45 feet above the ground. The timbers used in the construction of these bridges were all native Spruce. The aggregate length of the trestle work on this portion of the line was a little over one mile. Two hundred men were employed in the construction of the line, and received a wage of $8.00 a day. This was fifty cents more a day than the workers received when the first phase of the railway was constructed in 1905, and was considerably more than the wages paid to the laborers out in the creeks. At best, they could only command $6.00 a day plus board.

At approximately milepost 20, which was about two miles out from Fox and a short distance before Gilmore was reached on the original line, the railroad made a spectacular 180 degree reverse turn, over a curved wooden trestle. Later, under Alaska Railroad ownership, this trestle would be filled in with rock and gravel. At the north end of the "loop" stood the *Loop Roadhouse* which served the needs of the traveling public, both rail and otherwise. It was owned by Henry Baatz of Gilmore. Around December 27, 1909, it burned to the ground. Undaunted, Baatz rebuilt the structure — rising from its own ashes, Phoenix-like, within a month of its destruction. On Saturday, January 29, 1910, it re-opened in proper style with a grand party for the large crowd which was in attendance.

This rebuilt roadhouse served the public well until April 15, 1914, when it met a questionable and strangely coincidental demise — again by fire! Also burned to the ground, at this time, was the Gilmore railroad station. The origin of the fire was unknown, for it occurred in the early morning hours, with the result that there were too few men about to contain it. Some of the personal property was saved, but in actual fact, both structures represented a total loss. However, it was the strong contention of Henry Baatz, the owner of the roadhouse, that the fire originated from a spark which came from a passing Tanana Valley Railroad locomotive — which co-incidentally passed his establishment only a few minutes before the blaze was discovered. He stated that the fire broke out under the eaves of the roof near the corner of his bedroom; and since the fire originated more than 20 feet from his stove or chimney, it would have been impossible to consider either one of these sources as the culprit.

The fire at the railroad station was discovered shortly after 11 A.M., and within a very few minutes upward of 40 men had rushed to the scene to fight the flames. A valiant attempt was made to save the depot, but a strong wind was blowing which ultimately spread the flames so rapidly that it was a seething mass of red-orange in only a few minutes. Another structure which stood close by the depot was in danger of being destroyed, but it was drenched with gallons of hand-carried water, much to the relief of its owner, W.W. Archibald. No personal injuries were encountered during these ordeals, except for Henry Baatz, who sustained a cut and a few burns on his forearm.

Interestingly, an earlier move of the station from one location to another had already occurred before the fire. When the Loop trestle was constructed, the original track to Gilmore had to be pulled back one mile in order to allow the railroad to make the broad reverse curve across the valley to prepare it for the trek across the hills.

Now, much to the chagrin of the people of Gilmore, their station was not only a total loss, but there was a possibility that it would not be rebuilt. Originally it was the intention of the railroad to replace, as soon as possible, the destroyed depot with a new one. However, according to C.W. Joynt, General Manager of the railroad, this would not be the case. Instead a *boxcar* was positioned in the approximate location of the burnt station — to be used as the new depot! On the other hand, Henry Baatz never rebuilt his roadhouse, and would not allow anyone else to build on those charred premises. Instead, he opted to reestablish himself in Fairbanks and open a new business.

After passing the Loop Roadhouse, the railroad began a 2¼ percent climb up the side of the forest covered hill. At milepost 22 the railroad made a 90 degree turn to the north to parallel Fox Creek. Sitting precariously on a rocky, tree covered slope, the serpentine line wove along a shoulder cut from the side of the hill. This often left at least one side of the shoulder, or "cut", exposed to the sun — along with the underlying ice lenses. The heat from the sun changed the solid ice into a mass of water, which in turn flowed away carrying large amounts of soft ground with it. This resulted in a weakness of the roadbed requiring extensive repairs and deposition of gravel in these washed out areas. On at least one occasion, this damage to the roadbed caused the track to slip as the train passed over it, causing the train to careen off the track and plunge down the side of the hill.

Passing over many small wooden culverts and bridges the railroad approached, on a curve, while hugging a solid wall of rock, bridge No. 5. This was the first of two large trestles only a quarter of a mile apart. Trestle No. 6 which followed, also on a curve, was a spectacular edifice which was the high point of this spectacular section of the railroad. Starting at milepost 25, this trestle curved almost 180 degrees, over a distance of 600 feet, to place the track in a reverse direction appearing much like that of a hairpin. Quoting from a reporter during a railroad excursion to Ridgetop, he stated, "The most notable feature about the crowd (on the train) was the large number of cameras in evidence, which were brought into service every time the train swung onto one of the long curved trestles that marked the

Paragraph 4 on page 52 (which concludes on page 53) references Fairbanks Daily Times, April 16, 1914

The destruction by fire of the second Loop Roadhouse on April 15, 1914. It was located at the north end of a curved 403 foot wooden trestle at milepost 21.4 of the railroad.
Photo courtesy of the Candy Waugaman Collection

Author's Note:
The confirmed destruction by fire of the Loop Roadhouse puts to rest the supposition held by many local Fairbanks residents, that the Cleary Summit Ski Lodge was constructed from the timbers of this building.

Paragraph 1 references information from report by the Board of Road Commissioners of Alaska, 1908, pg. 18

Paragraph 2 refernces Fairbanks Daily Times, April 26, 1914

Paragraph 3 references information taken from a letter written by Falcon Joslin to Riggs, A.E.C., in September, 1915

Paragraph 4, page 53 (which concludes on page 54) references Fairbanks Daily Times, August 18. 1907

"My heart is warm with the friends I make,
And better friends I'll not be knowing;
Yet there isn't a train I wouldn't take,
No matter where it's going."

—Edna St. Vincent Millay

Paragraph 3 references Fairbanks Sunday Times, August 16, 1908

grade up around Fox. Superintendent Moriarty also came in for his share of snapshots, as he trailed behind the excursion train in his (rail) auto, ready to lend a hand if the train should snag itself or throw a (break) shoe".

Still climbing in this reversed direction, the line, after about a quarter of a mile, made another 180 degree turn to the north where it crossed Fox Creek for the last time, to breakout, at milepost 26.1, on to a bleak, wind blown plateau. On the north side of bridge No. 6, there still stands to this present day, three rough-hewed log section houses which in all probability belonged to the railroad. Growing directly out from one of the rotting timbers is a stately Birch tree, seemingly wanting to give everlasting life to these cabins.

Once the railroad broke out onto this windy terrain of stunted trees, it was for all intents and purposes, traveling the crest of the hills which acted as a barrier for a direct route to Chatanika. At milepost 26.1, the line passed through *Scrafford* at an elevation of 1,430 feet. A siding, 180 feet in length, was located here. It was named after Eugene L. Scrafford, a well-known prospector and miner in the Fairbanks region. Just a little farther down the line at milepost 28, the railroad crossed the highest point on the line, *Summit*, at an elevation of 1,496 feet.

One and one-half miles downgrade from Summit, was Ridgetop at milepost 29. Located north from here was the section of line where, in the winter, the railroad was subjected to arctic, gale force winds and deep, wind driven snow drifts which towered above the trains. In the summer, it became a sports and recreational mecca for the whole mining region. Trains would come up from Chena and Fairbanks loaded with merry makers clutching their picnic baskets in one hand and their berry picking pails in the other. It was a time when each outing and picnic would also feature a prime attraction — the baseball game! Where else could one witness such contests between two teams with such melodious, if not malodorous names, such as the *Fish Eaters* verses *The Roughnecks*, all for the price of $2.50 round trip, by train!

Ridgetop, reached by the railroad by August 10, 1907, was perched on the crest of a downward sloping grade at an elevation of 1,087 feet. Consisting of the depot, which contained the "town's" only telephone, and a few isolated businesses, it functioned as a distribution point for railroad delivered freight. These supplies were destined for *Dome City* which was situated east of the crest at the bottom of the hill, and to *Vault*, which was also located at the bottom of the hill, but on the west side of the crest. Freight was delivered by horse-drawn wagons over a steep and precipitous, winding road.

Founded in 1906, a year before the railroad reached Ridgetop, Dome City, which had no railroad service, was located on Dome Creek right opposite Discovery Claim. It could be reached two ways from Fairbanks; one could take the train to Fox and then transfer to Cleveland's Stages for the $5.00 one way trip, or one could come over the 18 mile trail by wagon, keeping in mind that it could be shut down at any time, even while in transit, by rain, mud or snow! On the other hand, the railhead at Ridgetop, although 29 miles from Fairbanks, was an all-weather route, capable of hauling the smallest or the greatest of loads, delayed at times, but never totally closed down its operations because of inclement weather.

Dome City and the surrounding mining region of Dome Creek had a population of 850 people. Besides the ever present saloons, it had a hotel with the sumptuous name of *The Grand*, a hospital, two cafes, six merchandise stores, two bath houses, a laundry and — two dance halls. It was served by telephone from Fairbanks and had electricity for the community. Law and order was preserved by the local *gendarme*, a U.S. Marshal. Today, not only is it physically gone, but worse still, history seems to have forgotten it.

Vault, founded in 1905-1906, was located at the confluence of Treasure Creek where it flowed into the northward bound Vault Creek. It was a much smaller community of about 250 people and consisting of the obligatory saloons, three to be exact, three general stores, and two log-hewn roadhouses. It still "lives" today, but in another respect. Close to the original site of Vault passes the world famous Alaska Oil Pipeline as it cuts its way through Alaska from Prudhoe Bay to Valdez. The access road which once lead to Vault still exists, as a turnoff from the Elliot Highway. The highway, on the other hand, pretty much parallels the old Tanana Valley Railroad's right-of-way from milepost 26.1 to milepost 29, and represents the only road to Alaska's Arctic.

Hugging the natural contours of the hill, the track, now surrounded by dominant Black Spruce trees, descended down a steep grade into the next valley. Turning north again after entering this valley, the railroad entered *Olnes*, which had a 205 foot siding. It was located at milepost 33.7 on the railroad, at an elevation of 609 feet. Named after a Norwegian immigrant, it too, acted as a transfer point for shipments destined for the mining camps at *Tolovana* and *Livengood*, which had come up from Chena and Fairbanks. At the same time, *Olnes* was the transfer point for heavy boxes of gold dust which were being sent out from these camps for railroad dispatch to Fairbanks. In later years, it became the transfer station for tungsten and antimony ore which also came out of the surrounding mining camps. This ore was shipped out by rail to the Stamp Mill in Chena, to be crushed into small particles in preparation for the removal of the minerals.

Olnes was a drab community with roads buried under a foot of dirty, smelly, black mud. The "town" consisted of one log restaurant, two saloons, the railroad depot, and — off on a railroad siding, a large combination warehouse and machine shop. The miner's homes were, in most cases, nothing more than

CHECHACOS **Should take a trip over the line.**

The view from the summit and the mining scenes onroute, have no equals in Alaska.

Tanana Valley Railroad Co.

THE TWO PASSENGER COACHES

ARRIVED LAST NIGHT

But the mail coach got lost in the woods. Scotty will bring it along if he finds it. In the meantime in place of letters from home you can buy yourself a nice book at the PIRATES. Also nice stuff for wearing apparel. Hosiery, Underwear, Cooking Utensils. We have no $10 bills hid among our goods. Our hungry creditors take care of these.

BUT EVERYTHING IS CHEAP ON THE

Pirate Ship

Eldorado water and Fire Department
Photo courtesy of the George Simpanen Collection

poorly built and poorly insulated, rough-hewn log cabins. As a result, they could be so cold that in the winter, the miners would have to wear their same woolen clothes both inside and outside the cabin.

Generally, if these miners had wives or girlfriends, they were not to be found in the community, for with one exception, there were no other girls in "town". This one exception was the cook that prepared the meals for the miners. During the week she would live out this lowly role, to become on Saturday nights, the only girl in town, the "Belle of the Ball"! It was customary on this evening, to "strike up the band" and trip the light fantastic on the dance floor. The "band", in this case, was a willing miner who would invert a wash tub and beat out a "rhythm". There upon, the miners would line up, and individually ask for the lady's hand for a promenade across the mess hall floor!

Dante's Inferno: A book of prose describing passage through hell, "La divina commedia" by Italian poet, Alighieri Dante. (1265 -1321)

One such lady was Lois McGarvey, who arrived in town by train, as the only non-male passenger on board. The men in the passenger car, upon her entrance, greeted her with scorn and implied derision — and indicated to her that they would have been much happier if she had stayed off the train. She described the scene as if it came straight out of *Dante's Inferno* — "They were all drunk, or nursing bad hang-overs." The seats faced each other, two and two, leaving her no option but to sit face to face with two sleeping drunks. Several men were smoking strong cigars, and as the foul and nauseating smoke permeated the car, she became deathly sick. In attempt to keep her stomach contents in their proper place, she lost control of the rest of her body with the result that she simply flopped around on the cushion of her seat, while her hair bobbed about synchronously with the rock and roll of the train.

Paragraphs 1-3 reference
Along Alaska Trails, pg. 21

Besides running a boarding house for the miners, she worked as the station mistress at the railroad depot. Lots of honest miners patronized her establishment, but with the fluctuations of the fortune of life, they could not always pay for their room and board. To re-coup her financial losses, she made a deal with the General Manager of the railroad, Mr. Joynt, whereby she would obtain the concession to supply the fuel for the wood-burning locomotives. After much deliberation, he agreed. There-upon, she summoned all the willing debtors, and directed them to the task of cutting, sorting and stacking cord wood for eventual use by the railroad. In return she was able to collect, vicariously, all money owed to her by the miners — from the railroad!

Leaving Olnes on a 6 degree curve to the east, the railroad passed through a heavy growth of mixed forest. Approximately 150 yards into this grove the railroad passed over a small brook, which was of great importance to the line for west bound trains. In order for the train to get over the hills to Gilmore and beyond, it needed a full head of steam. Since this brook was the only source of water between these two locations, a water tank was built at this site. The brook still flows to this day. Ultimately, this water tank would meet its fatal demise in a strangely symbolic manner. On April Fools day, 1920, the crew of the snow train wrote an ignominious chapter into the scrolls of the Tanana Valley Railroad. On this day, which was bitterly cold, the crew prepared the snow train for a trip up the hill to Ridgetop. During the previous two days, one of the worst storms in the history of the camp had been raging up and down the line. Drifts between Olnes and Chatanika were plowed clear time and time again, only to fill up almost immediately after the snow train had passed through. The snow became so crusted and hard, that upon being struck by the charging snowplow, it did not scatter into a powdery cloud, but flew apart into solid, fragmented pieces of compacted ice and snow!

Locomotive No. 50 stopped to take on water at the typical water tank. Note the spout visible above the train and the link and pin coupler on the front of the locomotive.
Photo credit: University of Alaska Fairbanks, Alaska and Polar Regions Department Acc. #79-41-55, in the archives, of the Falcon Joslin Collection, Album #2

In preparation for another charge up the hill, the engineer of the snow train slowly and deliberately notched his throttle up with each spin of the wheels, to get the momentum and speed necessary to part those walls of snow. Snorting white steam from the cylinders, while simultaneously pouring out bellowing clouds of black smoke and sparks from its balloon stack, the train grimly raced past the water tower in anticipation of its charge up the hill. Just as it passed the water tower, the structure burst into a mass of flames. Red hot sparks, emanating from the locomotive stack, showered the timbers of the tank, which had not been used for a long time, nor had seen a drop of water for years. It burned completely to the ground!

Paragraph 2 references Fairbanks Daily News-Miner, April 2, 1920

Shortly thereafter, the line broke out onto a broad valley of tundra and stunted boreal forest. Roughly following the Chatanika River, the line, almost in a perfect tangent, went off to the northeast as it passed through this muskeg, simultaneously skirting the fringes of many small ponds. At milepost 35.7, almost at its final destination of Chatanika, the railroad came into *Little Eldorado*. At the time that the railroad initially passed through this location, it was known by a variety of names, such as White's Siding, Eldorado City or by whatever name a miner wanted to call it. Originally, it was nothing more than a mossy stretch of tundra where a merchant by the name of J.L. White, had the foresight to establish a modest general merchandising business. With the strike of gold along the immediate creeks, new buildings went up so fast that *hourly* bulletins had to be posted to let the local residents know what was going on. In all this haste, flatcars arriving with lumber for this construction were emptied so quickly for an immediate turn-around to Chena or Fairbanks, that the load was just dropped to the ground where the railroad car was spotted. Main street, which paralleled the track, had such a construction boom that it was soon realized that the regular train service, even with added cars, could not meet the needs of this new bonanza. To meet these demands, the railroad added an extra, or "special" train to its schedule whose sole purpose was to deliver trainload freight and other merchandise directly to Little Eldorado. White's Siding was to be the railroad transfer point at which supplies would be unloaded and forwarded up the creek to the miners at their claims.

Paragraph 1 & 2 references Fairbanks Daily News, October 18, 1908

Eldorado Trading Co.

LITTLE ELDORADO

All Kinds of Lumber

Fairbanks Daily News

Author's Historic Note:
This structure has been erroneously considered by some Fairbanks residents as the old Eldorado railroad station. It is not! The actual station is pictured on the cover of this book.

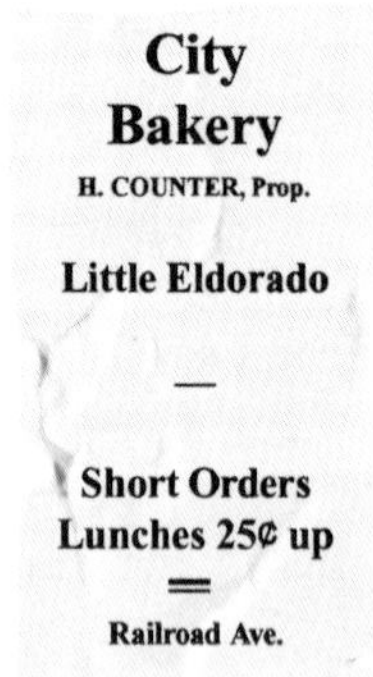
City Bakery

H. COUNTER, Prop.

Little Eldorado

Short Orders
Lunches 25¢ up

Railroad Ave.

Fairbanks Daily News

Paragraph 4 references Tanana Valley Annual Report, 1908

As far as the town citizens were concerned, there was a desire to make it into a modern city with all the up-to-date utilities. They scorned the idea of using crude kerosene lamps. Instead, they were electrically wiring up the town to connect with the Tanana Electric Company lines so as to be able to burn incandescent lights, and *run coffee-grinders with electric motors*!

So fast and so great was the need for housing that the new Eldorado Hotel, which was formally known as the Chatanika Hotel and stood some 12 miles farther down the Chatanika River, was brought up *piecemeal* to Little Eldorado. In less time than it took to find the logs in the woods, Howard Barnes, the owner of the structure, sub-divided the old building into selected pieces and shipped these units, in carload lots, up the railroad line to its new location to be reassembled into the original hotel. Ironically, long after the hey-day of Little Eldorado, an attempt was made to salvage one of the hotels from this community. It was dragged en-masse, back up along the Chatanika River to a new location where it was to be used as a hunting lodge. This move fell short of its intended destination, and so today it stands, leaning at a precarious angle along side of the abandoned roadbed of the railroad — rotting away as it draws closer to its ultimate demise. Whether or not these two hotels were one and the same is unkown!

Leaving Little Eldorado, the railroad, still in its initial tangent, passed through a tundra-like terrain, until milepost 38 was reached. Here the line took a 4 degree deviation to the south after crossing *Ruby Creek* at milepost 38.1. To this day, one is still amazed by the sight to behold after Ruby Creek is crossed. On the north side of the roadbed mining operations appear to be frozen in time. Tools are scattered about; living quarters and a steam generating plant with a resident boiler intact, are still standing. Close by rests a wood pile which may well have been cut and stacked many years ago when the wood was still green — but now lays rotting and falling apart. Amidst all of this is a "Gin Pole", appearing ready for use as it did almost three quarters of a century ago! This is the remains of the *Mad Russian Mine* waiting patiently for its people to come back!

After the railroad passed Ruby Creek, the line entered into a one mile long reversed curve to finally come to the end of its destination at milepost 39.2 — *Chatanika*! On September 29, 1907, the *Last Spike* was driven into the waiting tie to mark the completion of the ***Tanana Valley Railroad***.

Initially known as 15 below - *Cleary*, Chatanika started out as a mining camp in which tents were the basic housing available to the miners. Settled in 1904 on the west bank of Cleary Creek, which was near the Chatanika River, it assumed its prominence in the region in 1907 when the larger nearby community of Cleary City suffered a disastrous fire. Half the city was destroyed. Instead of rebuilding, half of its population of 750 people elected to move to Chatanika. Subsequent to the arrival of the Tanana Valley Railroad to this mining camp, at about the same time that it was bolstered with an influx of new miners coming from the ashes of Cleary City, Chatanika evolved into a terminal center for the railroad, a distribution point for freight destined to the surrounding regions, and a home for about 500 people.

Chatanika had a wood framed depot and freight house, and most of the usual amenities typical of these mining communities. Since Chatanika was the end of the line, reversal of the trains was accomplished by a 1,734 foot loop of track which was located approximately nine-tenths of a mile west of the station. The recently arrived train would back down the track to enter the loop, which then curved back to the mainline and re-entered the station, but now in a reversed position — ready for its return back to Fairbanks.

Some of the other buildings in town were of similar frame construction to the station's, but a greater portion of the structures consisted of rough-hewn logs, chinked with moss. A post office was opened in Chatanika on March 20, 1908, only to close temporarily on November 30, 1910. It continued this periodic cycle of opening and closing right up to the demise of this town. Although its population gradually declined after 1909, Chatanika did not pass into history through abandonment and decay; rather, it was actually dug-up and cast aside into multiple mounds of rocks in the latter part of the 1920's by the dredges of the Fairbanks Exploration Company in their quest for gold!

Main Street, Little Eldorado
Photo courtesy of the George Simpanen Collection

Fairbanks Daily Times, Wednesday, September 11, 1907

THE GRAND HOTEL

14 Single and Double Rooms ∴ Mrs. Persons ∴ LITTLE ELDORADO

European Plan - Modern in Every Respect
Electric Light - Telephone

Particular attention is given the Dining Room and Everything the Market affords is served

Well Stocked Bar in Connection

Fairbanks Daily News

Tanana Valley Railroad Co.

TIME TABLE NO.10
Effective 12:01 a.m., June 8, 1908.
SUBJECT TO CHANGE WITHOUT NOTICE.

No. 5 Mixed	CHENA AND FAIRBANKS	No. 6 Mixed
8:00 a.m. Leave	Chena	Arrive 6:00 p.m.
8:25 a.m. Leave	Junction	Leave 5:30 p.m.
8:45 a.m. Arrive	Fairbanks	Leave 5:15 p.m.

FAIRBANKS AND CHATANIKA

— Northbound —			— Southbound —	
No. 1 Passenger Daily.	No. 3 Mixed Ex. Sunday.		No. 4 Mixed Ex. Sunday.	No. 2 Passenger Daily.
Lv. 9:30 am	Lv. 1:15 pm	Fairbanks	Ar. 11:40 am	Ar. 4:50 pm
9:45 am	1:30 pm	Junction	11:20 am	4:35 pm
9:50 am	1:40 pm	Ester	11:00 am	4:30 pm
Meet 10:25 am	2:20 pm	Eldorado	Meet 10:25 am	4:05 pm
10:55 am	2:55 pm	Fox	9:50 am	3:45 pm
11:15 am	Meet 3:30 pm	Gilmore	9:30 am	Meet 3:30 pm
11:55 am	4:15 pm	Ridgetop	8:45 am	2:50 pm
12:20 pm	4:50 pm	Olnes	8:05 am	2:20 pm
12:45 pm	5:25 pm	Chatanika	7:30 am	1:45 pm

Trains No. 1 and No. 2 have right of track over No. 3 and No. 4
FALCON JOSLIN, Pres. and Gen. Mgr. A.P. TYBON, Supt.

Fairbanks Daily News, June 21, 1908

N.C.CO.
MACHINERY
July 10
1911

CHAPTER 7

A Tale Of Two Cities

Fairbanks, Alaska, July 10, 1911
Photo courtesy of the Nicholas Deely Collection

The history of the origins of Fairbanks is really a tale of two cities, that of itself, and that of its earlier downriver rival, *Chena*. Indeed, they both were called, at the time of their inception, *Chenoa*! The crucible of their origins was gold! It wasn't that either one or the other had any gold strikes, but rather, it centered about the concept as to which one of the communities was in the most favorable position to serve the miners. Since the primary mode of transportation during this period was river travel, both fought for this preeminence.

Author's Historical Note:
Chenoa, in the Athabascan Indian tongue, means "River of Rocks". At a later date, the spelling of the town was arbitrarily changed to Chena, due to the absence of the letter "o" at the printer's shop.

Chena's laurels rested on the fact that it sat directly on the banks of a navigable river, the Tanana, near its confluence with the Chena, where water levels remained adequate at all times to accommodate shipping. Fairbanks' claim to be that of a transportation hub was that it, too, rested on the banks of the navigable Chena River and that it was six miles farther upstream and, therefore, closer to the mines. This proximity would make freight and transportation charges much cheaper for the trek out to the creeks. So sure was Fairbanks that the Chena River would remain a navigable waterway, that it built its own boat building facilities in 1905, which predated Chena's shipyard by three years. However, travel up the Chena River was not without its problems. Periodically, because of low water which was actually a recurrent problem, one of the riverboats would get stuck on one of the many sandbars which plagued the river. Larger companies, like the North American Transportation and Trading Company, and the Northern Navigation Company, would charge these stranded boats $100 per hour to be winched off the sandbar. It is interesting to note that when the steamboat *Koyukuk* arrived in the Interior in 1906 carrying locomotive No. 52, a 2-6-0 Mogul, some railroad cars and rail for the Tanana Mines Railway, plus six stage coaches for the Fairbanks-Valdez winter trail, it came up the Chena River to Fairbanks, by-passing its own yards in Chena. Locomotive No. 52 was obtained second hand from the Denver and Rio Grande Western Railroad and worked the line with its three other engines right up to the demise of the Tanana Valley Railroad.

Steamboats On The Chena, pg. 65

Steamboats On The Chena, pg. 67

By 1907, when the second segment of the Tanana Valley Railroad was nearing completion, Fairbanks had grown from a river port of 300 people in 1903, into a sophisticated community of 5,000 inhabitants locally, with another 5,000 people scattered throughout the mining towns and creeks. By far, this was the largest population center in Alaska. This phenomenal growth continued to parallel the fortunes of the mining industry, right up to the period of its greatest productivity in 1909. The subsequent fall in mining activity did indeed cause a slow and gradual decline of Fairbanks, but unlike other mining communities all across North America, it kept on rising, again and again, like a Phoenix rising from its own ashes. Whether through just plain luck or defiant intent, Fairbanks has continued to survive and to grow as an academic, cultural, military and transportation center. However, the basis of this success was formulated right from the very inception of the community. Principally, it was evident that the town could not rely alone on the mining industry, but rather a diversification of the economic base had to develop. To maintain the flow of people and materials, an effective transportation network was

Locomotive No. 152, purchased by the Alaskan Engineering Commission for use on the Tanana Valley lines and the Nenana extension. It was the last and only new engine that the system had. It exists today actively working on a tourist railroad in Michigan. Notice the coal in the locomotive tender. Of all the engines on the Tanana Valley lines, only this locomotive and No. 151 utilized coal.
Photo credit: Anchorage Museum of History and Art, Acc. #BL 79.2.5010

needed. This came, initially, through a co-ordinated water transportation system from the sea to the inland rivers, to be supplemented later, with a network of overland distribution in the form of rail and highway arteries. Next, it developed an effective political institution which had the power to operate and finance public utilities. Lastly, the town developed the conveniences, comforts, entertainment and social institutions which were in concert with other established communities.

The Development of Settlement in the Fairbanks Area, pg. 47

It is rather ironic that this town, having secured its destiny through the political machinations of Judge Wickersham, should have ultimately grown to such a size; that in the end it was to physically engulf its old nemesis, Chena. Perhaps arguably, Chena should have been the place to survive over Fairbanks, for it had at its doorstep a truly navigable river and a railroad for overland transportation, which would have made Fairbanks' proximity to the mines insignificant. Beyond its initial role as a transfer junction, that is, from boat to rail transportation, no other community structure ever was to develop. In time it was to be regulated to that of an historical oddity, doomed to a role of virtual obscurity.

Panoramic view of Fairbanks, July 10, 1911. Notice the riverboat on the Chena.
Photo courtesy of the Nicholas Deely Collection

On the other hand, if one was going to Alaska, Fairbanks was the place to go. Arriving on one of the river steamers, especially during sunset, after having passed through hundreds of miles of wilderness, one could not be but awed by the sight which one was to behold. Like an over-zealous Christmas decoration, electric lights were glittering throughout the community, permeating the darkness of the wilderness with light. In conjunction with the magic of electricity, telephone and telegraph had arrived shortly after its inception as an organized community. In actual fact, the telegraph system which was to link it to the rest of the world, arrived in Fairbanks in 1903, the same year that the town was to be incorporated as a city.

In the downtown core area, steam heat, as well as a water plant, was available for the business establishments. In the winter, however, running water was not available, but was delivered in five gallon containers which were distributed to subscribers from horse-drawn sleighs. To keep the water from freezing into solid ice, charcoal heaters were placed on board the sleighs whereby it was possible to keep the water in a fluid state.

Water being delivered in the winter to customers throughout the Fairbanks area. Note the charcoal heater.
Photo courtesy of the Candy Waugaman Collection

Yet Fairbanks still was a frontier town. Although it had two hospitals, St. Joseph's Catholic and St. Matthew's Episcopal, with eleven physicians and six dentists, every other building seemingly appeared to be a *Saloon*! However, from this hodge-podge of buildings, there emerged a panoply of fine shops, ice cream parlors, indoor ice and roller rinks, and most importantly, a public swimming bath. These were all laced together with a rough-hewn board sidewalk which at times could make for delicate walking. Not all boards were firmly attached to the stringers, so that if one stepped off center on one of these boards, it would "jump up" and solidly slap one down!

Along Alaska Trails, pg. 18

Paragraph 1 & 2 references to *Along Alaska Trails*, pg. 18 & 19

An advertisement for Barthel Beer carried in the 1907 Tanana Valley Directory.

THE BEER THAT WILL MAKE FAIRBANKS FAMOUS

Times Force of Experts Receive Life Size Sample of Bartel's Best Home Product to Sample.

Schlitz beer made Milwaukee famous. The first keg that was brewed was sent to a newspaper office and the staff joined in giving the beer such a good name that its fame spread rapidly.

It might just as well be announced now that it will be Bartel's brewery that will make Fairbanks famous for a keg of the first lager, with all its purity, tone, taste and body reached the Times office last evening.

"Vas is?"

The Times force acknowledges the corn malt and declares positively that all other beers are mere imitations.

Fairbanks Daily Times, August 22, 1906.

Paragraph 4 references Fairbanks Daily Times, August 22, 1906

Paragraph 5 references Tanana Valley Directory — 1907 & Fairbanks Daily News-Miner, December 18, 1915

Also, as it is with all frontier towns, the males far exceeded the females. So bad was the situation, that when one heard the deep "too-too-too" from the whistle of an arriving steamboat, the whole population of the town, including a *year old calf moose*, came hurrying down to the dock. The prospectors' interest centered around the cargo being brought in — but equally important, any young female passengers that might be coming in from the outside.

The year old calf moose was a favorite of the Fairbanks people and ran around the town as if he owned it. Everyone gave way whenever he came swinging along the narrow sidewalk. He paid no attention to anyone, for he was usually *bound for one of his favorite bars*, where the miners would *let him drink from their glasses*. Finally, he became too domineering and, eventually, was led back to the woods.

Perhaps more than any other institution, the *Saloon* has become synonymous with the frontier, and as such, this town of 5,000 adventurous souls, had at least 15 "mug-up" havens. If SCHLITZ® was the beer that supposedly made Milwaukee famous, then it was felt that Bartel's Brewery, Fairbanks' very own, if for no other reason than the amount consumed, would make this town famous.

The first keg of beer that was brewed was usually sent to the offices of the local news media, the Fairbanks Daily Times, where it was later reported that, "The staff of experts joined together in giving the beer such a good name, that its fame was to spread rapidly." It spread such joy among the staff that after having a fair sample of the brew, they claimed, "That with all its purity, tone, taste and body, all other beers were mere imitations." Not to be outdone in the attempt to quench the parched tongues of the miners and to dull the realities of life, Tanana Brewing Company was also on the scene to help the miners part with their gold dust. Once primed, these lads could strut their stuff at the local dance hall or if in extreme "distress", one might find their way to the infamous 4th Avenue to relieve one's tensions at one of the string of "houses of joy". Sunday, if so inclined or needed, they could repent at one of Fairbanks' four churches.

It was a town that developed character and self-respect whereby it could withstand both the ravages of God and man. On July 4, 1905, the Chena River, which periodically was plagued by low water and uncertain navigability, changed from a tranquilly flowing body of water into a raging torrent, spilling over its banks and inundating the city. Within a year, on May 22, 1906, a fire broke out which destroyed the whole business district within 42 minutes. If that wasn't enough to discourage the towns' people, then the infestation of the community by *Rats* in 1915 could break the spirit of all but the heartiest. These rats came into the Interior as "stowaways" on board the river steamers that came in from St. Michael where contact was made with ocean-going vessels. What became of them is uncertain, but one could assume that it was a bumper year for both Fox and Lynx.

The fire occurred at 3 o'clock on the afternoon of the 22nd of May. It was first noticed by the wife of Dr. Moore, a Dentist, when smoke came rolling down the walls of the Washington-Alaska Building, the location of his office. In all probability the whole town, with its collection of dry wooden buildings, would have vanished had it not been for the presence of an existing fire department, a wind which sprang up from the south and forced the fire to the water front, and the creativeness of the young manager of the Northern Commercial Company. As a result of this set of circumstances, the burned area was confined to the district bounded by Turner, Lacey, First and Third Avenues. With the exception of the Fairbanks Banking Company's building and the warehouse in the rear of it, nothing was left standing in the four great blocks which comprised the commercial heart of Fairbanks.

Fairbanks Times, May 23, 1906

What spurred on the action of the manager of the Northern Commercial Company was the deplorable sequence of events that were occurring as the fire progressed. Initially, the usual large supply of cord wood cached out in Weeks Field was depleted. This supply was used by the city power plant, which was operated by the Northern Commercial Company, to pump out water to the mains. As the fire increased in severity and spread over a greater area, hoses from the fire department became burned and leaked like sieves. All this occurred at the same time that additional water mains were being opened up as the fire increasingly engulfed more and more of the city. Because of the lack of wood to fire the boilers, water pressure gradually diminished to the point that no forcible spray could be maintained.

Author's Note:

The heat value of a log of dry Birch depends on the concentration of woody material, resin, water and ash. When wood is compared to fossil fuels, a full cord of dry Birch will weigh approximately two tons and equals (approximately) the heating value of one ton of hard coal or 130 gallons of fuel oil. Generally one cord of dry Birch will provide 18,200,000 BTU. (British Thermal Units)

One cord of dry Spruce will provide approximately 15,000,000 BTU.

One ton of bacon would provide approximately 28,000,000 BTU.

(14,000 - 15,000 BTU / lb.)

Being knowledgeable of the incendiary quality of fat, the manager of the Northern Commercial Company put out a call for all the bacon that was available in the store, while at the same time, requesting the town folks to gather up all the bacon that they could muster at home. Like chariots in the Roman Coliseum madly dashing about, one can almost imagine how these teams came storming through the streets of Fairbanks with their precious cargo of six and ten pound slabs of bacon as they charged over to the power plant. With their cargo delivered, these same wagon drivers then joined in with the others to stoke the boilers. By the time that the last ember faded into history at 5 P.M., 2,000 pounds of bacon were thrown into the boilers. "Bacon, the basic food of any Alaska camp, was worth 40 cents a pound over the counter, cash or credit. In the boilers of the Fairbanks Power Plant, it was worth four-fifths of the city of Fairbanks."

Flag Over the North, pg. 2

No fatalities were reported during this most disastrous fire and the town was rebuilt without any increase in prices by the merchants, the lumber company, or the Northern Commercial Company. This Company which was an off-shoot of the Russian-America Company, became established in Fairbanks in 1903 and was destined to play a very significant role in the interior of Alaska. Though best known as a mercantile chain and operator of a power plant in Fairbanks, it also played a role, to some degree, in the field of transportation.

Flag Over the North, pg. 3

Fairbanks Daily News, July 17, 1907

On July 16, 1907, the city council of Fairbanks passed, for the first reading, an ordinance which granted the right to the Northern Commercial Company to erect and maintain a tramway along Third Avenue to the city limits. This line was built to the same three foot gauge as the Tanana Valley Railroad, and was extended out to Weeks Field. This field served as a storage area for wood that was to be used in the power plant of the Northern Commercial Company. Thus, the tramway would serve as a transportation facility to haul this wood to the plant.

Wood consumption at the plant was at least 40 cords of wood a day, and up to 15,000 cords a year. It was cut to four foot lengths and would usually arrive in the winter on sleighs stacked sky high, which were pulled by tremendously powerful horses. It was then stacked row upon row, over extensive acreage at this wood reservoir. In order to maintain a steady flow of wood for its insatiable furnaces which were some distance into town, some sort of a "conveyer belt" was needed. This proved to be the new tramway that was being built, which was quite often referred to as the Northern Commercial Railroad. It has been postulated that some of the track on this pike was the redundant line that had been left behind by the Tanana Mines Railway during the futile period when extension out to the Sourdough District was contemplated. The wood and other supplies were hauled on flatcars similar to those used on the Tanana Mines Railway to the power plant, and to their machine shop on Third and Barnette Streets.

Weeks Field submerged in wood for use by the Northern Commercial Company power plant. Evident in the foreground is a track belonging to the "Northern Commercial Railroad". In the background is the Chena River Bridge in Fairbanks.
Photo credit: University of Alaska Fairbanks, Alaska and Polar Regions Department Acc. #73-66-124, in the archives, of the Charles Bunnell Photograph Collection

Fairbanks Daily News-Miner, February 20, 1915

No locomotive was ever owned by Northern Commercial Company. The employees of this small pike were particularly proud of their railroad and would boast that "It was a perfectly good railroad, always behaved itself, never ran over anybody, or smashed anybody's hands in the couplers, and doesn't scare the town with whistling, or cover it with smoke from the engine house — because it owns no engine!" They were very sensitive, however, to the fact that although the pike was a railway in its own right, it was the Tanana Valley Railroad that had the community's attention. Their frustration was being voiced in that, at the time when they had relaid *20 feet* of track on the town's railroad, the only work that had been done on it for years, there wasn't so much as a bit mention of the improvement in the newspaper.

Although this pike was primarily an industrial railway, it did, once a year, become a common carrier — destined to perform a very vital civic and holiday function. The 4th of July in those days was truly a very big national holiday. Parties, dances, parades and a myriad of social affairs occurred. But the big event for the day was — *Baseball*! Teams would come into town from as far away as Ridgetop and all the surrounding camps. Since the picnic and the ball game were held in Weeks Field, which was some distance from town, the "railroad" would transport these sports fans and merry makers from the power house of the Northern Commercial Company out to the ball field. They put together some seats fashioned from crude cut lumber, placed them on the flatcars, and hauled the crowd out to the ballfield with horses and mules. Since the game and the picnics were full of "spirits" in every respect, the railroad provided an additional service by safely bringing back to town these "sports fans". This time however, they would just stack them up width wise across the flatcar like cords of wood. This service was discontinued in 1924 when the steam plant was converted to coal which was now available to the community from the Healy coal mines.

By 1906, the Federal Government awarded a contract to the Northern Commercial Company to haul mail, passengers and freight down the 376 mile winter run to Valdez from Fairbanks. Initially, mail going to the "outside" was carried by individuals who for one reason or the other were leaving Alaska. For a pinch of gold dust, one would take the mail to a costal town, such as Skagway, and post it. For a few extra pinches of gold dust, it would be carried all the way to the states by the traveller and posted there.

For passengers going to Valdez from Fairbanks, the fare was $125 — but for one coming north from Valdez, the fare was $150. The outgoing freight was most often gold which had come in from the mines out in the creeks. From 1905, when the trains started to run out to the creeks, each arrival would bring back its new batch of "millionaires" — and sadly, a new batch of paupers! Nearly every train had among its passengers four or five travellers with a singular purpose. Tucked away in unimpressive bags or *deep* pockets, were buckskin pouches of gold dust that only a few hours earlier were embedded in the soil out in the creeks. There was no outward appearance of any riches that might be riding the rails at any one time, but common knowledge led the Tanana Valley Railroad to name its priority train the "Gold Dust Limited".

More than seven million dollars in gold bullion was carried for 12 years by the Northern Commercial Company's sleighs over and across the mountains to Valdez — without any loss. This movement was carried out without the benefit of either armed guards or fire-arms for the drivers. In actual fact, it was not the use of fire-arms which represented the source of social disorder during this period, but rather the free flowing use of alcohol. The presence of guns in the general populace was at a ratio of only one to five hundred.

This respect for law and order put the Fairbanks jail, at one time, in a precarious position. The population in the Federal Jail, in 1920, had *decreased* alarmingly. The work of the reformers, or something else, had put the camp (Fairbanks) in such a shape that, work as they would, it be almost impossible for the Deputy Marshal or the court to keep the jail properly filled. It was feared that, "If the situation keeps up and gets worse, it may result in the Department of Justice in Washington ordering the jail closed. This would throw a lot of people out of work and employment, and be a sad blow to the young life of our camp and society, in general!"

U.S. MAIL CONTRACT FOR CARRYING MAILS
MAIL ROUTES:

Contract Terms: — July 1, 1906, to June 30, 1910, Northern Commercial Company has following contracts:

From Valdez to Fairbanks, 345 miles, and back twice each week from Nov. 1 to April 30, and once a week during October and May in each year. Carrying any class of mail the department may elect, the total weight not to exceed 800 pounds per single trip each way. N.C. Co. 519 Crossley building, San Francisco, Col. $64,368. 10 days for trip, with 5 days additional allowed in October and May. Bond, $70,000.

From Circle to Fairbanks, via Mastodon: — 100 miles, and back, twice a month from Nov. 1, to April 30, in each year, carrying any class of mail the department may elect, the total not to exceed 300 pounds per trip each way. 6 days allowed for the trip each way. $6,000 bond.

From Fairbanks to Tanana, via Chena and Tolovana: — 275 miles, and back, once a week from October 1 to May 31, in each year, any class to be carried that the department may elect: the total weight not to exceed 800 pounds per single trip. 3 days allowed for each trip, with an additional 4 days in May and October, when necessary. $25,000 bond.

From Tanana to Fairbanks, via Chena: --275 miles, and back, six times a month in safe and suitable steamboats, and as much oftener as contractors may run, from June 1 to September 30 in each year, supplying all intermediate points along the Tanana River: trips to be about 6 days apart. Pursers on steamboats to act as railway post office clerks without additional compensation. $10,000 bond.

Contractors must be of legal age and carriers must be over 15 years old.

U.S. Mail Contract For Carrying Mails, as recorded in the 1907, Tanana Valley Directory, pages 25 & 26

Paragraph 2 references Fairbanks Daily News-Miner, June 19, 1908

Flag Over The North, pg. 308

Fairbanks Daily News-Miner, February 16, 1920

Gold Rush Town, pg. 24

Paragraph 2 references All Alaska Weekly, July 23, 1971

Paragraph 3 references Fairbanks Daily News, August 28, 1928

By the same token, the concept of a jail in the early days of Fairbanks, was entirely different from our present day conceptions. In spite of having 31 attorneys to manage the "social ills" of the community, convicted criminals were treated on a friendly, *Do-As-You-Please* basis. As an example, Circle City actually had a sign hung out in front of the jail which read, "All prisoners must report by 9 o'clock, P.M., or they will be locked up for the night!" — by order of the U.S. Marshal. Apparently, they had no reason to try to escape for they had a warm room, food and were hundreds of miles from nowhere.

While Fairbanks grew haphazardly, Chena was, from the onset, laid out in a grid a half mile long, in which many of the streets were named after U.S. Navy battleships to commemorate them for their service in the Spanish-American War. So strong was the conviction of the speculators in Chena that it would become the heart of the Interior, that virtually overnight this bit of riverside wilderness was transformed into a town which was well organized and designed. It had a hospital, telephone system, school house, several newspapers, a police department, a jail, as well as a 500 seat public amusement hall.

BEER SMUGGLER WAS DISCOVERED

An organized system of illicit smuggling on the island waters of our beautiful city has been discovered and the Federal authorities have just taken steps to stop it.

For some time past a mysterious individual has been seen tacking up Second avenue several times daily under full sail and carrying a heavy cargoe which he stored in a suit case.

The apparent weight of the contents of the hold seemed to drag him down below the water line, and the observant ones became obsessed with the idea that he was carrying gold dust or had made a steamheated touch in a junk shop.

These strange journeys continued with such regularity and such mystery that they finally became the subject of judicial investigation with the result that it was ascertained that the heavy cargoes were liquid and consisted of the stuff that made Milwaukee the home of the palm garden.

Further inquiry developed the fact the bottled hops were purchased at a retail bar in what constituted wholesale quantities and it is more than likely that the traffic will experience a sudden sensation.

Fairbanks Daily News, August 28, 1928

The flaw in this whole design, however, was that in actual fact Chena was a "company" town. "Company," in this regard, refers not to a corporate entity, but rather to individual men who fostered the whole plan for their own personal financial gain. People like Martin Harrais, who initially was so involved with the construction of the Tanana Mines Railway, became very instrumental in the development and construction of the town. It has been reported that he spent $86,000 of his own money, which represented an enormous financial outlay at this time, for the construction of service industries and other dwellings. He built, owned, and operated the Chena Lumber and Light Company, in conjunction with a saw mill, which furnished the lumber needed for the construction of the town's dwellings. In 1908, during the decline of Chena, he built for Schubach and Hamilton, a river shipping company, the largest shipyard in the Interior. This company operated the river steamers *Monarch* and *Julia B.* between St. Michael and Fairbanks, which connected at the former port with ocean going vessels. In time, Martin Harrais was to become the first mayor of Chena and, later, a delegate on the Democratic ticket to a territorial convention in Valdez.

In its hey-day, 1904-1905, Chena had a population almost as large as that of Fairbanks. By 1907, it was very much on a downward slide with only about 450 people still living in town. From 1906 to 1909, when both Fairbanks and the Tanana Mines Railway were booming, Chena managed to hold on, somewhat, although this was in actual fact, the beginning of the end for the river port. Never-the-less, Chena was not totally doomed. The Tanana Mines Railway established its main offices, yards and repair and car-building shops here. Because of this, the dock was kept busy with the result that these operations were able to sustain some of the few remaining businesses in town.

Strangely, though originally the railway had sustained the town, it was to become a big factor in its demise. The management of the line, soon realized that the shallowness of the Chena River was to be less of an obstacle to river navigation than was initially expected. With Fairbanks fast becoming a cosmopolitan center, the choice port of call for these steamers was to be this city in preference to Chena. Consequently, incoming freight that arrived by river steamer to Fairbanks, would go directly to the mining creeks and Chatanika, by-passing Chena altogether! As a result, there was very little freight going from Chena directly to the mines. Instead, it was first routed to Fairbanks, from Chena, and then on to the mines. In time it was to be referred to, with some derision, as the "Old Dummey Line", as romanticized by this old ballad:

VALE CHENA! On the old dummey line.
So, we left Fairbanks at half-past one
and got into Chena at the setting of the sun.
On the old dummey line!

In time there developed a custom in Chena, that whenever there was a new bride on the train, the locomotive would let off a continuous, melodious blast of the whistle whenever it came into the depot. However, on one particular night there was something peculiar about the long bellow of the locomotive whistle that, ultimately, attracted more than the usual number of "gawkers". Speculation was rife as to which one of the young maidens on the train was to be the bride this time. Men, women, and children rushed off in great haste to a nearby bridge which was located by the depot. Leaving the train first was a group of workmen followed by a woman or two, who the "gawkers" felt were plainly not in the bride class. In the distance they caught sight of a yellow canary cage bobbing up and down. The woman holding the cage was too far away for those on the bridge to distinguish her features — nor was it certain that she was dressed in the regulation white. As happens with crowd hysteria, in which their imaginations run wild, they started to shout, "That's her, that's her!" As this mature woman approached the crowd, with a big self-satisfied smile on her face, she slyly responded with an explanation as to the nature of the whistle blowing, — "There were a couple of jack-asses on the tracks, and *there appears to be a few of their near relatives on the bridge*!"

Four-wheeled electric car parked in the Chena yards. Photo credit: University of Alaska Fairbanks, Alaska and Polar Regions Department Acc. #91-46-821, in the archives, of the Frederick B. Drane Photograph Collection

Fairbanks Daily News-Miner, April 16, 1920

Author's Note:
By 1919 the Tanana Valley Railroad was quite affectionately referred to as the *"Old Reliable Tanana Valley Railroad,"* while that portion of the line that went from Chena Junction to Chena, for some reason or another was called the *"Dummey Line."*

Fairbanks Evening News, August 24, 1905

ELDORADO TRADING CO
GENERAL MERCHANDISE.
BANK
STORE
POOL
RESTAURANT

CHAPTER 8

THE GOLDEN YEARS 1907 - 1909

The station stop at Little Eldorado. Note the unique water tank in the distance and the hand car on the siding.
Photo courtesy of the Dory Stucky Collection

Link and pin couplers. **Photos courtesy of the Candy Waugaman Collection**

With the completion of the railroad to Chatanika in 1907, an unprecedented boom occurred in the mining operations of the region. Freight, particularly heavy mining equipment and passenger loads, increased dramatically. The Tanana Valley Railroad, now 44.7 miles in length, with an added three miles of sidings, became the final conduit in a vast transportation network that consisted of ocean-going vessels, riverboats, overland wagons and, in the winter, sleighs. Ultimately, trucks and automobiles would be competing for the lucrative business out to the creeks.

Overview of Chatanika, the modest railroad yards with the engine house off to the left, and the mining enclaves in the distance.
Photo credit: University of Alaska Fairbanks, Alaska and Polar Regions Department Acc. #73-79-47N, in the archives, of the Tony Troseth Photograph Collection

The presence of the railroad resolved, for the interior of Alaska, the issue of exorbitant cost and uncertainty related to the movement of heavy boilers and other equipment from navigable rivers to inland destinations. This was, perhaps, the biggest singular obstacle that existed to the growth and development of the mining operations. The expense involved in the movement of freight by wagon or sleigh *(if it could be moved at all)* ranged from one dollar a ton per mile using a "first class" wagon road, to ten dollars a ton per mile via trails and streams. For this reason, anything of mining value that was 50 miles or so beyond an established transportation mode, in spite of its potential value, was left untouched!

Paragraph 2 references *Alaska - Yukon Magazine*, pg. 247, January, 1909

Once the trains started to travel out to the creeks, 1905 freight rates dropped to 86 cents per ton mile. By 1908, this rate decreased even further to 58 cents per ton mile. Passenger rates too were appreciably reduced — travel came down to 13 cents per mile! While still more expensive than stateside rates, the railroad saved the people of Fairbanks approximately $300,000 a year in freight costs. By the same token, between January 1907 and September 1908 the railroad spent $500,000 for wages, averaging $25,000 per month. This included the salaries paid to the construction gangs as well as to the regular employees. In addition, all supplies that were needed for this construction were purchased, as much as possible, from the local merchants. As a result, a large percentage of these expenditures were put into the economy of the community.

Chatanika bound mixed train crossing one of the many trestles located between Fox and Ridgetop. In all probability this is either bridge No. 5 or No. 6, located at milepost 24.6 or 25.1, respectively.
Photo courtesy of the Candy Waugaman Collection

Paragraph 3 references Fairbanks Sunday Times, October 18, 1908

Total construction costs for the Tanana Valley Railroad amounted to $867,000. This represented the sum total of expenditures incurred for equipment, surveys, transportation of materials coming up from the states *(everything except the timbers and ties used in the construction of the railroad)*, and actual construction costs including excavation and labor. Of this sum, $300,000 went into the second construction phase, that is, the line from Gilmore to Chatanika. Joslin's original estimate for the extension of the railway was $107,000. In spite of this added cost, however, it was a virtual necessity that it be built, for it tapped additional mining areas that were devoid of any adequate transportation services. The whole railroad was built without any government subsidy except for a five year dispensation from a federal tax of $100 a year for each mile of track.

Paragraph 4 references Alaskan Engineering Commission Report, March 12, 1914 to December 1915, Tanana Valley Railroad Annual Report, 1908 and *Ghost Railway in Alaska*, pg. 9

During the year of 1909, the year of its highest earnings, the Tanana Valley Railroad tallied a gross of $298,250.77. Its expenses were $182,347.77 leaving it with a net earning of $115,902.77. Thus, its

Billing courtesy of the Dick Farris Collection

Paragraph 1 - 3 references Fairbanks Daily News, Ticket rates-December 12, 1907; and an Article from July 27, 1908

operating expense, exclusive of interest, was about 62 percent of earnings. However, from the standpoint of an investment, the Tanana Valley Railroad never proved to be a financial success. It was able, never-the-less, to meet expenses and pay interest on its indebtedness of $666,000.

As might be expected, freight tonnage reached its greatest peak during 1909 when the railroad hauled 15,809 tons, surpassed only in the year of 1912 when it hauled 16,842 tons. Passenger volume was equally robust during this period, transporting 49,205 travellers. Special fares were periodically offered that represented a substantial reduction from the usual rate. A return ticket to the following destinations was:

Chena	$2.00
Ester	$2.00
Big Eldorado	$2.50
Fox	$3.00
Gilmore	$3.50
Ridgetop	$4.75
Olnes	$6.00
Chatanika	$7.00

New Rates

The Tanana Valley Railroad Co.

THE TANANA VALLEY RAILROAD CO. HAS ISSUED THEIR NEW FREIGHT TARIFF COVERING WINTER RATES TO ALL POINTS — PROMPT SERVICE IS GIVEN. SHIP YOUR FREIGHT BY RAILROAD AND SAVE TIME AND MONEY — WARM CARS ON EVERY TRAIN FOR PERISHABLES. FOR COPY OF TARIFF AND ALL INFORMATION APPLY TO ANY OF OUR AGENTS OR

J.H. ROGERS,
TraMo Manager

All this increased mining activity during these "Golden Years" called for more and more freight and passengers to be transported out to the creeks. But, with this increase in transportation came an increase in the maintenance of the line. Ironically, at the same time that seemingly unlimited amounts of freight were being landed at Chena, its waterfront dock was on the verge of collapse. The Tanana River had been cutting away the shore line, and with it the docks which served the railroad. Had not immediate repairs been made, the whole town of Chena might have suffered even a much earlier demise.

No sooner had it been announced that the dock was ready for service on July 26, 1908, that riverboats started to arrive at such a rate that the town of Chena was now buried under massive amounts of cargo. The North American Transportation and Trading Company riverboat, *J.P. Light,* was the first to arrive on the following day with two barges in tow carrying 800 tons of cargo. At the same time, there were many other riverboats strung out along the Tanana and Yukon Rivers heading for Chena with hundreds of tons of freight. These included the steamers *Barr* and *Cudahy* which were already at the mouth of the Tanana River with additional tons of freight. Far down the Yukon River at its mouth on the Bering Sea, St. Michael, the transfer point from ocean-going vessels to riverboats, was becoming inundated with more and more freight — unable to move the load until the riverboats made their return trip.

To accommodate all this freight and to get it out to the creeks as quickly as possible, the Tanana Valley Railroad had two trains running all summer long over the entire system. Increasingly larger and larger boilers were coming in on the river traffic. Since they were desperately needed by the miners to thaw out the permafrost during the short summers of the Interior, they had to get out to the creeks as soon as possible. These steam boilers were used to thaw the ground in the drifts, excavating as much as 400 cubic feet a day. To meet these demands, more of the larger and heavier equipment was coming into the mining sites. As a result there was an increase in the development of these operations. While a large amount of these supplies were actually going to Fairbanks, most of the freight was going over the divide to Dome, Vault and the Cleary creeks where there was an increase in mining development.

Paragraph 1 referernces Fairbanks Daily News, July 26, 1908

All of this heavy train movement was going on at approximately the same time that the line was upgrading its track. Every spring, the breakup would do tremendous damage and destruction to the track and bridges. Trains would actually stop, while in transit, to allow for the immediate repair of the track. In many instances, this meant that the track was actually being elevated above its usual level to counteract the catastrophic, seasonal flooding which destroyed both track and bridges.

Mixed freight train preparing to leave Chatanika for Fairbanks on a cold winter day. The picture must have been taken after the Tanana Valley Railroad came under the wing of the Alaskan Engineering Commission, which is confirmed by the presence of an automatic coupler on the end of the flatcar. Prior to this period, the Tanana Valley Railroad used link and pin couplers exclusively. Photo credit: University of Alaska Fairbanks, Alaska and Polar Regions Department Acc. # 79-41-74, in the archives, of the Falcon Joslin Photograph Album #2

Soon after the *Cudahy* came in with its 800 tons of freight, four other rivercraft arrived with a total of 2,100 tons of cargo, and with more coming in! Besides the *Cudahy*, which had been unloading its freight for several days, the *Healy* with 450 tons, the *Barr* with 400 tons, and the *Monarch* with 450 tons arrived almost simultaneously on August 2, 1908. So tremendous was the deluge of cargo, that General Agent Tyson got out from behind his desk to direct the movement of freight out on the docks. With only three locomotives at his disposal, the fourth being the small Porter 0-4-0 that was limited to yard service and the run to Fairbanks, he ran special trains, almost continuously, hoping for a quick turn around from the creeks in order to make the most efficient use of the limited rolling stock.

As summer turned into fall and the cold temperatures settled in, there developed by the railroad management a great concern for the perishables. Not to be out done by the weather, the railroad installed small stoves in some of the boxcars and redesignated them as "hot cars!" Perishables could now travel to their destinations in relative safety from freezing. Approximately 50 years later, the Bangor and Aroostook Railroad of northern Maine adopted this same technique as a means of safely shipping their potatoes to market, all over the eastern states.

High up in the hills between Fox and Olnes, the mixed passenger train has stopped in the vicinity of Summit. Note the two "hot" cars trailing the locomotive. Photo credit: University of Alaska Fairbanks, Alaska and Polar Regions Department Acc. #79-41-40, in the archives, of the Falcon Joslin Photograph Collection

Although the movement of freight took priority over most other things, track maintenance and the upkeep of the superstructures on the line were equally paramount. In 1908 an engine house was constructed in Gilmore to provide shelter for a locomotive and its crew so that they could remain in a state of readiness to combat, when called upon, the phenomenal snow accumulations expected over the divide at any time during the winter. It was their intention to keep the track clear at all costs, so as to

Fairbanks Sunday Times,
October 18, 1908

Traffic jam at the railroad station in Little Eldorado. Mogul No. 52 in the background.
Photo courtesy of the Candy Waugaman Collection

Fairbanks Daily News-Miner,
April 24, 1909

Fuel stop in Fairbanks, 1913. Wooden sides were added to the tender of engine No. 51, a 2-6-0 Mogul, to increase the tender's total capacity.
Photo courtesy of the Candy Waugaman Collection

All-Alaska Weekly *(Also known as Pioneer All-Alaska Weekly)*, April 20, 1923

maintain an uninterrupted flow of passengers and freight out to the creeks along the way to Chatanika. During this same year, the railroad completed arrangements for horse-drawn teams to connect with the railroad at various points along the track so as to facilitate the delivery of freight from these locations out to the miners, who could be several miles away from the point of transfer.

With the onset of spring 1909, it became evident that freight and passenger loads out to the creeks would continue to be heavy. On April 24, 1909, the number of trains out to the creeks was increased to three a day to meet this increased demand for rail service. At 9:30 A.M. on that day, a mixed train left Fairbanks for Chatanika, followed at 11:30 A.M. by the *"Goldstream Special"*, a train which went to Gilmore and back. This was a direct run, not stopping, as was the usual procedure, to load up with wood. This quick turnaround was necessary, for the train had to be back in Fairbanks in time to become the 4:00 P.M. train to Chatanika. This, too, was a direct run to Chatanika, not stopping anywhere along the line to load up with additional wood. It was found that by increasing the height of the sides of the locomotive tender with wooden siding, greater loads of fuel, in this case wood, could be carried. Thus, the tedious and time consuming stops enroute to the train's destination were eliminated, thus allowing the locomotive and cars to be recycled for additional runs during the day.

Short trains on the Tanana Valley Railroad were the rule and not the exception. While generally speaking this represented no problem to the movement of large amounts of freight, it did call for much ingenuity in the attempt to transport certain types of cargo. Some of these problems were caused by Interstate Commerce Commission regulations which were not truly adaptable to the operations of the Tanana Valley Railroad. Since train lengths very seldom exceeded six or seven cars, the prohibition of the transport of dynamite to no closer than ten car lengths from the detonating caps made the movement of this much needed explosive to the mining camps a virtual impossibility. Further, under no circumstances could the dynamite or the detonating caps be carried in a passenger car.

To skirt these regulations, the freight agent would select an innocent passenger and ask the traveler if he would be so kind as to carry a "special railroad package" aboard the train with his own personal belongings. In the meantime, the dynamite was placed in a boxcar. Upon arriving at the designated destination for the dynamite, the passenger was asked to pass the bundle on to the conductor. Except for the freight agent, everyone involved in this charade was innocent. While this delivery was being carried out, another employee of the railroad would fetch the dynamite from the boxcar never once realizing that the I.C.C. regulations were being violated.

Just as inventive in their attempts to solve difficult problems, was the railroad's approach to handling unique freight loads. On one occasion, just as the train was pulling out from Fairbanks, Art Hering of Sourdough Express, a company that still exists today, delivered three pieces of one inch galvanized pipe, twenty feet long, for delivery to the mining camps in Chatanika. Although the train was in a hurry to depart, a futile attempt was made to get the pipes into a boxcar whose dimensions were shorter than the length of the pipe. At long last the train pulled out of Fairbanks to arrive, later in the day, in Chatanika. The truck from the Fairbanks Exploration Company was there to meet the train, but the pipes were nowhere to be found. Even after being reassured by the Fairbanks dispatcher that the pipes were on board, the train crew, even after repeated search, could not find them. The Fairbanks dispatcher remained adamant that the missing freight was on board, in spite of the continued verbal abuse, in retribution, by the train crew. On the following day, when two boxcars were being switched onto a siding, a brakeman up on the roof of one of the cars noticed, as he was setting the brakes, the missing pipes tied to the roof walk. Apparently, the agent in Fairbanks had ordered the pipes to be attached to the roof of the boxcar without notifying the train crew.

Tanana Valley Narrow Gauge,
Frontier Times, July, 1969

Tanana Valley Railroad Company

Time Table No. 13. Effective 12:01 a. m., May 10, 1909

This Company reserves the right to vary from this Time Table at Pleasure

Chena and Fairbanks

North Bound Daily No 7 Mixed		South Bound Daily No 8 Mixed
Lv. 7:00 a. m.	Chena	Ar. 7:00 p. m.
Lv. 7:25 a. m.	Junction	Lv. 6:25 p. m.
Ar. 7:50 a. m.	Fairbanks	Lv. 6:00 p. m.

Fairbanks. Gilmore and Chatanika.

North Bound				South Bound		
NO. 1 Mixed Daily	NO. 3 Goldstream Special Daily	NO. 5 Chatanika Passenger Daily		NO. 6 Fairbanks Passenger Daily	NO. 4 Goldstream Special Daily	NO. 2 Mixed Daily
A. M.	A. M.	P. M.		A. M.	P. M.	P. M.
9:30 Lv	11:09 Lv	3:40 Lv	Fairbanks	10:00 Ar	2:30 Ar	5:30 Ar
9:45 M	11:15 Lv	3:55 Lv	Junction	9:45 Lv	2:15 Lv	5:10 Lv
9:50 Lv	11:20 Lv	4:00 Lv	Ester	9:40 Lv	2:10 Lv	5:00 Lv
10:25 Lv	11:50 Lv	4:30 M	Big Eldorado	9:10 Lv	1:40 Lv	4:30 M
11:00 Lv	12:20pm	4:55 Lv	Fox	8:45 Lv	1:10 Lv	3:50 Lv
11:20 Lv	12:30 Ar	5:10 Lv	Gilmore	8:30 Lv	1:00 Lv	3:35 Lv
12:10pm		5:55 Lv	Ridgetop	7:50 Lv		2:50 Lv
12:40 Lv		6:20 Lv	Olnes	7:25 Lv		2:20 Lv
1:00 Lv		6:35 Lv	Little Eldorado	7:10 Lv		2:00 Lv
1:10 Ar		6:45 Ar	Chatanika	7:00 Lv		1:50 Lv

North bound Trains have right of track over South bound Trains.

FALCON JOBLIN, President — A.P. TYSON, Gen. Mgr.

Passengers holding tickets to points beyond Gilmore on trains No. 1 and 3 may stop over at Fox or Gilmore to train No. 5 on date of sale only; and passengers on train No. 6 holding tickets to Fairbanks may stop off at Fox or Gilmore until train No. 4 or No. 2 on date of sale only.

School "train", Fox, Alaska, sometime during the late 1920's.
Photo courtesy of the Olga Steger Collection

Mixed passenger train at Ridgetop Station, at milepost 29, and at an elevation of 1087 feet. Since there was no real community there, the sleighs off to the left of the boxcar were probably used for service down the hill to Dome and Vault.
Photo credit: University of Alaska Fairbanks, Alaska and Polar Regions Department Acc. #79-41-73, in the archives, of the Falcon Joslin Photograph Collection

Paragraph 1 references Fairbanks Daily Times, November 7,1909

Excursion train preparing to depart from the Fairbanks depot. The crowd is so large that the railroad has had to press into use its open air "smoking cars."
Photo credit: University of Alaska Fairbanks, Alaska and Polar Regions Department Acc. #77-89-23, in the archives, of the Vide Bartlett Photograph Collection

Fairbanks Daily Times, May 31,1906

"Smoking Cars"
Photo courtesy of the Candy Waugaman Collection

Fairbanks Daily News-Miner, April 17,1909

Fairbanks Daily News-Miner, April 17, 1909

In time, it was possible to make direct connection to an off line destination through the use of ancillary services. Passengers going to Ester, for example, would take the electric car to Ester Siding and then transfer to *Parker's Auto* for the completion of the trip. Sadly, no sooner had this service with the electric car commenced, than a entrepreneur by the name of *David Courtemanche* established an *auto* service to Ester City. This was a daily service leaving at 9:00 A.M. from the *Nordale Hotel.*

While the major part of the railroad's revenue came from its transportation of freight out to the creeks, passenger service, from two aspects, represented another source of substantial income. Regular passenger service was usually of the "mixed" variety. That is, to the rear of a freight train was coupled a single coach, or more if needed. The train, therefore, served two functions at the same time. However, there was another source of passenger revenue — one of which has become very popular to this present day — *excursions.* Movement of crowds of people during holiday and other social events put a maximum strain on the four coaches of the Tanana Valley Railroad. To compensate for this periodic great influx of passengers, the railroad altered some flatcars and designated them as "Smoking Cars!" These cars were not constructed primarily for males to enjoy their cigars, but rather were just plain flatcars with added plank benches. Smoking car, in this case, meant that the passengers riding this car were to be exposed to the smoke coming from the locomotive! Never-the-less, "It was a great place to hear discussions on the mining industry and many a bucket was hoisted, and many a hole put to bedrock between Fairbanks and Gilmore!"

These excursions took a multitude of directions, and for relatively small communities, the Fairbanks people, the Chena folks, and those out in the creeks did a lot of railroad socializing. Such was the case when the Fairbanks Arctic Brotherhood went, en masse, to Chena on April 17, 1909, to visit their brethren in that community. "Welcome To Our City" was plastered all over the town, while the mayor and the locals, in-spite of the supposed mutual animosity that existed between the two communities, went out to meet, with cheers, the train as it pulled into town. With the "Golden Key" in hand, and the band playing to their heart's content, the Chena folks met the wild and enthusiastic bunch from Fairbanks — both parties ready and anxious to start the festivities with a rousing opening of the Masquerade Ball! All were assured by the mayor that there would be "something stirring every minute" and that, "anyone receiving blurred impressions, during the course of the evening, need not necessarily hurry to an oculist!"

As the train prepared to leave Fairbanks at 6:00 P.M., for the 45-50 minute run to Chena, there was such a throng of people taking this train that the added use of the "smoking cars" was contemplated, — and used! This excursion train was given the right of way, "with orders that freight and passenger trains, president's specials, and hospital conveyances, would be run in the clear. So special were these runs *(and all passenger excursions)* that the conductors were admonished not to "spit between the vestibules of the cars," and that they must "wear their shoes at all times"! These instructions were not only verbal, but were written into the operating manual of the railroad!

Not to be outdone by their Chena friends, the Arctic Brotherhood of Fairbanks staged their own Masquerade Ball at the local roller rink the following New Year's Eve. Because of the enormity of the projected crowd, one could purchase a ticket only after receiving a special invitation. People from Vault, Dome, Cleary and Chatanika came in on the regular train, while the folks from Ester and Chena came into Fairbanks by a special run. Both trains departed from their respective stations early on the evening of the ball. The trains would make the return trip whenever the ball was over, regardless of the hour. Tickets were $3.50 for the gentlemen who, in turn, were permitted to bring *one* lady with them. Should they desire to bring an *Extra Lady*, it would cost them an additional $1.00 for each lady, to cover the cost for dinner, which was customarily served after the ball! No one was allowed in without some kind of a costume, and spectators were absolutely forbidden.

Railroad excursions were also very popular during the 4th of July festivities. This was a period in our American history, when the holiday was celebrated in a grand old style with picnics, outings and the "big baseball game(s)." So feverish was the public for these games, that the enthusiasm began to cut into the profits of the local saloons — better known as the *"Thirst Emporium."* Advertised in the local newspapers were such admonitions as, "Don't spend all your money for baseball — have a high ball at The Fraction."

Since Fairbanks was noted for its frequency of parties and social affairs that were open to the folks from all around the region, the creek "towns", on July 4, 1907, reciprocated with an invite to celebrate the holiday with them. Virtually the whole town of Fairbanks went off into the hills and creeks with booze, food, baskets and baseball bats! The 9:30 A.M. train proved to be insufficient to carry the crowd out into the hills, so an additional train was put on the schedule for 11:30 A.M. The fare was $2.50 which entitled the ticket holder to many benefits. Since the activities were strung out along the entire length of the railroad, one could travel all the way to Chatanika and back, with the right to stop off at any one of the intermediate destinations. Further, the price of the ticket entitled the holder to be admitted to the "big" baseball game which was to be played later that afternoon in *Ridgetop*!

Another really big annual event of the year in Fairbanks was the *Fair* held late in the summer or in the early fall. As it is today, farmers grew fresh vegetables for the local market in 1909, culminating, at the end of the season, with the annual Fair. People would come into town from all the surrounding areas. As a result the railroad was hard pressed to put all its coaches into service to bring these crowds in from the creeks. Excursion rates for round trips were in effect during this period which acted as a further inducement to come to the Fair. The load usually consisted of two coaches from Chatanika and another

Author's Note:
From a conversation with the late Robert Turnbull, who stated to me that he saw these admonitions, with his own eyes, in the railroad's timetable/manual of the era.

Paragraph 1 references Fairbanks Daily News-Miner, December 30, 1909

Little Eldorado Station is on the left of the photograph. Notice the handcar on the siding, and the unique water tank in the distance.
Photo courtesy of the Dory Stucky Collection

Paragraph 2 references Fairbanks Daily Times, August 4, 1908

Paragraph 3 references Fairbanks Daily News-Miner, June 24, 1907

Fairbanks bound mixed freight train stopping for water at Olnes before climbing the hill to Ridgetop.
Photo courtesy of the Candy Waugaman Collection

one from Gilmore. Where roads were open to the mines, automobiles weighed down to the axles with their large human load started to compete for this mass of humanity that was on their way to the Fair. This interest in the Fair was so great that the railroad, in spite of making every concession possible, did not have enough coaches to transport all the folks to the Fair. During this period in the history of Fairbanks, the Fair was held in a hall that also acted as a roller skating rink during the rest of the year. At the close of the Fair, there was a big celebration and dance before the creek people returned back to their mines to sit or work out the long cold winter.

Fairbanks Daily News-Miner, September 5, 1911

All of this heavy train movement did not go on, however, without invoking Murphy's Law — that is, *if anything can go wrong, it will*! On Monday, June 22, 1908, the rear passenger car of a Tanana Valley Railroad train jumped the track near Olnes and did a somersault in mid-air. It came to rest with its roof on the ground, and its wheels pointed to the sky. Only two people were in the coach, and it was reported that they *enjoyed the novel experience*! After a short period of time and effort the car was replaced on the track. No damage was done except for a few scratches on the car. So quick was the response to the accident by the railroad work crew that there was actually only a slight delay in the service.

Fairbanks Daily News-Miner, June 23, 1908

~

Indeed, this time frame, 1907-1909, proved to be the best of times for the railroad, for it provided the miners with an economical and expeditious conduit for the delivery of mining machinery and supplies, impossible before its completion. Mining camp after mining camp now became strung out — not only along the right-of-way, but in many cases, at some distance from the railhead. Cleary, Ester and the Fairbanks Creeks in conjunction with their tributaries were prime examples of large mining operations off the main line — and yet, they were the streams that were responsible for making millionaires out of paupers.

A railroad wreck somewhere along the Tanana Valley lines of the Alaska Railroad. A caboose appears to have been present on this train, a seldom confirmed entity. It appears to have been rebuilt from a passenger coach. Date of the accident is unknown, but must have occurred sometime between 1924 and 1930.
Photo courtesy of the Jim Brown Collection

The railroad made it possible to use heavy mining equipment which could now be transported out to the digging sites. This heavy equipment was needed because the mining carried out in this region was not a *placer* operation, as was carried out at Dome Creek, the Klondike, or along the beaches of Nome. Rather, it was *hard rock mining* which forced the miners to dig down to bedrock before actual gold extraction could begin.

Initially, mining was, at best, a crude operation. To reach bedrock, the miner would first have to thaw out the ground. This was done by gathering up some newspaper or a kerosene soaked rag, which was then mixed with some wood. This was ignited to produce heat which would defrost the ground to a depth of about one foot. This softened soil was then removed, and the process was repeated again, and again, until the desired depth was reached to harvest the gold-bearing rock. These vertical shafts that went down to bedrock were between seven to eight feet square, and were dug to depths of 300 feet or more! The average depth, however, was approximately 50 feet. *Drifts*, or right angle tunnels to a descending shaft, were excavated to variable distances, reaching as much as 600 feet from the shaft. Usually however, the distances from the shaft averaged 50 to 100 feet. They were approximately six feet wide and six feet high, and were illuminated by candles. The roof was braced, every six feet, with heavy timbers. Every foot of dirt in this operation had to be thawed, dug out by hand, and hauled in a wheelbarrow to the vertical shaft, where it was raised to the surface by a "bucket"! In time, so many of these tunnels had been excavated, that in a cross section, they looked like the spokes of a wheel radiating out from its hub.

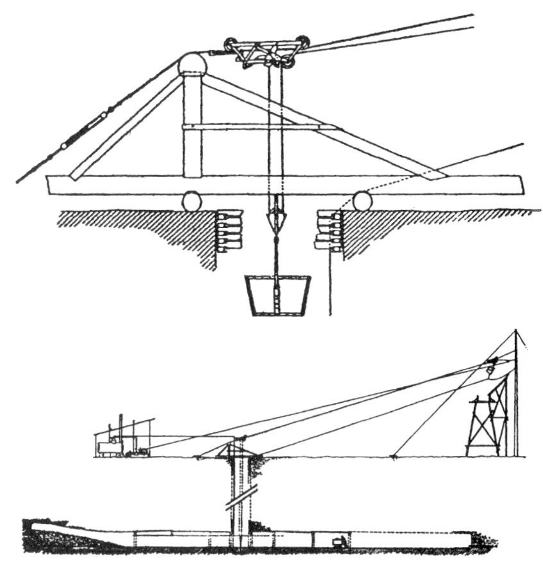

At the top of the shaft stood a tall vertical "tower" that was known as a *Gin Pole*. Attached to the "pole" was a system of cables, one of which was attached to the "bucket" that was raised and lowered down the bottom of the shaft. The miner would stand on the "bucket" firmly grasping the steel cable, and was lowered by a co-worker down to the bottom of the pit. These same "buckets" were used to bring the gold bearing rock to the surface for processing. The rock was then flushed through a sluice box, pending the availability of adequate amounts of water, whereby the gold would be separated from the rock.

Initially, this process of raising and lowering the buckets was carried out by a hand-cranked windlass. With the arrival of the railroad, heavier tonnage could now be hauled out to the mines which meant that larger boilers could be put to use. Their primary use was to provide steam or hot water streams which was used to thaw out the bedrock so that the ore could be harvested. With the boilers topside burning insatiable quantities of scarce wood to produce steam or hot water, the men below in the drifts hammered steam points into the frozen ground to pry loose the bedrock. Hot water was delivered, on the other hand, through cotton hoses, which had fireman nozzles attached to them, designed to shoot streams of hot water on the surfaces to be thawed. The hot water was preferred because it would soften only the ground to be mined, and leave the dirt above the miner's head, solidly frozen. Steam, on the other hand, fogged up the whole operation, softening the roof which made the danger of a cave-in ever present.

Unfortunately, great danger was a fact of life with this occupation. Miners would periodically fall down the shafts and suffer fatally by having their necks broken. Cables would, on occasion, snap and whiplash through the air to sever an arm, a leg, or cause severe maiming. In some cases, death was a welcomed visitor to this carnage. When these catastrophes occurred, the railroad would immediately dispatch one of its "speeders" to pick up the injured person for a priority run back to Fairbanks for emergency care at St. Joseph's hospital.

Author's Note:
A "speeder" is a small, four-wheeled, track vehicle powered by a gasoline engine.

Bags of antimony sit on a siding in Little Eldorado.
Photo courtesy of the Candy Waugaman Collection

Gold Production in the Fairbanks Area 1903 - 1923

Year	Dollar value of gold
1903	$ 40,000
1904	400,000
1905	6,000,000
1906	9,320,000
1907	7,845,000
1908	9,180,000
1909	9,650,000
1910	6,242,000
1911	4,608,000
1912	4,371,000
1913	3,677,000
1914	2,739,000
1915	2,638,000
1916	1,839,000
1917	1,358,000
1918	827,000
1919	772,000
1920	600,000
1921	608,000
1922	747,000
1923	628,000

Source: Robert L. Monahan, who extracted the data from U.S. Department of the Interior, *Mineral Industry of Alaska*, U.S. Geological Survey, various bulletins, 1902 - 1944.

The following table shows a condensed statement of earnings, expenses and traffic from 1909 through 1914:

	Gross Earnings	Expenses	Net Earnings	Number of Passengers Carried	Number of Tons of Freight
1909	$ 298,250.54	$ 182,347.77	$ 115,902.77	49,205	15,809
1910	250,537.30	167,440.79	83,096.51	53,248	15,535
1911	160,659.98	116,615.32	44,044.66	34,629	11,867
1912	186,868.00	100,702.01	86,165.99	38,915	16,842
1913	167,758.67	110,608.50	57,150.17	41,682	13,210
1914	116,579.50	94,259.81	22,319.69	27,832	10,231

The Tanana district alone produced $9,174,617.40 of the whole production in Alaska:

From Yukon River Points and Tributaries	$ 1,150,935.60
From Nome and Seward Peninsula	7,322,447.00
From Southeastern Alaska	4,000,000.00
Total	$21,648,000.00

Discovery Dates and Prospector's Name of Key Tanana Valley Mines

Cleary Creek	Dome Creek	Fairbanks Creek	Little Eldorado	Vault Creek	Smallwood Creek	Engineer Creek	Ester Creek	Steel Creek	Tenderfoot Creek	Chatanika Flats
Aug. 1, 1902	Sept. 11, 1902	Sept.12, 1902	Sept. 14, 1902	Sept. 30, 1902	Dec. 17, 1902	Dec. 29, 1902	Feb. 14, 1903	March. 25, 1903	Feb. 23, 1905	1906
Felix Pedro	D.A. Shea & Pedro Creek Sept. 11, 1902 Felix Pedro	D.A. McCarty & Goldstream Sept.12, 1902 Felix Pedro	J.A. Mathieson	T.M. Gillmore	J.C. McDonald	James Dodson	L.A. Jones	Karl Westiwich	E.H. Luce	John Dobbins

CHAPTER 9

OPERATIONAL PROBLEMS

Mogul No. 52 stuck in the snow high up on the crest of the hills near Summit. Although the Railroad had a snowplow, it was evident that only men with shovels were going to be able to free the train.
Photo courtesy of the Candy Waugaman Collection.

Winter snow has stalled Mogul No. 51 and its train somewhere high up on the crest of the hills. Snow, wind and cold were ever present in the hills, at anytime of the year.
Photo courtesy of the Candy Waugaman Collection

With great fluctuations of temperatures such as exist in Alaska, anything that is of a mechanical nature represents a potential problem, and an operational hazard. Alaska's environment is in a state of constant change to the extremes, which places a perpetual stress on man and his gadgets. Consequently, the railroad was forced to operate under conditions of constant adjustment, with continuous corrections of the variable environmental insults. Winter, however, presented the greatest emotional and physical stress, for it taxed both man and machine to the limit.

Almost immediately when the cold wind from the north strikes, it becomes evident that what should have been done yesterday cannot be carried out today. Any moisture in the air coalesces into ice crystals which clog up any nook or cranny. As a result, anything of a mechanical nature is now at the mercy of these ice crystals, gripping to a standstill, anything in its path with the tenacity of the talons of a Hawk. Perhaps this is why old-time Alaskans refer to the onslaught of winter's bitter cold blast as the "Return of the Hawk!"

Snow, particularly in large amounts, is known to cause its own set of problems. Even in small amounts, it has its own insidious way of causing extensive railroad disruptions. Each winter, these problems became very much evident, particularly on that stretch of track between Summit and Ridgetop. Small flakes of snow, as they were jostled and blown about by the wind, gradually became pellets of snow, and developed static electricity. When these small, "electrified" pellets started to drift from the force of the wind, some of them found catch places and immediately began to "weld" themselves into an innocent snow mass, in and around the rails and their joints. As a consequence, these hardened masses caused the locomotives and their cars to "ride the rails"; that is, they lifted them right off the track which would then cause a derailment!

Climbing a 2.3 percent grade towards Summit and Ridgetop, Mogul No. 52 needs the assistance of another locomotive to get through the heavy snows up on the crest of the hills between Fox and Olnes.
Photo courtesy of the Candy Waugaman Collection

It was evident, therefore, at the onset that since this line was to be an all season and all weather route, a snowplow was an absolute necessity. The shops at Chena took flatcar No. 57, braced it up with heavy wooden beams, and applied a wedge plow — that is, a "V" shaped, horizontal attachment — to the front of it. Then, to each side of this unit, close to the front of the car, "wing plows" were applied. These "wings" were extended whenever snow accumulated to a marked degree along the sides of the plowed track. To propel this unit, one of the locomotives was needed. It would, very slowly, pull up to the back of the plow and very gingerly approximate the couplers. Since, at this time, the old *"link and pin"* couplers were still in vogue, it represented a very dangerous operation. The brakeman actually had to hold, *with his fingers*, these drawbars (links) while he dropped, with his other hand, the pin in between the links of the plow and the locomotive. This was done while the locomotive was actually moving in on the car at the same time that the brakeman's fingers were still poised *in between the cars*! Whenever you saw a railroad man with his fingers missing, you were almost always assured that he was a brakeman.

At the appropriate location where the line was closed by heavy snow or a drift, this locomotive-snowplow combination would position itself an appropriate distance from the obstruction on the track. Then the locomotive would build up a big head of steam and slowly, with an increasing syncopation of its driving wheels, fiercely head down the track. With a great crescendo it would slam into the snow, cleaving it into massive waves of white powder, and clouding the sky with falling flakes. At times, the track was cleared with the initial attempt, but quite often that first try just represented the beginning of a prolonged ordeal!

Once the track was cleared, a moat-like trough was created in which the resulting snowbanks would continuously keep caving in because of the heavy accumulation of snow. To combat this situation, the plow would extend its "wings" and, as the unit went down the tracks, it would clear the upper elevations of the snowbanks, casting snow still farther to the side of the right-of-way. Periodically, however, the top-heavy snow would cave in *after* the plow had passed through, thereby trapping this whole unit so that it was no longer operational. Ironically, the rescuer had to be rescued!

After the "dozer" had gone through to clear the track, shovelers were put to work to complete the task. This crew is working in the Chatanika Flats.
Photo courtesy of the Candy Waugaman Collection

Sometime before 1918, this plow was converted into a *dozer*. The reconstruction took place in the Alaskan Engineering Commission shops at Garden Island. A structure similar to a boxcar was placed upon the original snowplow which thus enclosed the whole deck of the flatcar. It afforded a greater amount of protection from the winter elements, and an added measure of safety for the crew who manned the plow. Their job was to make periodic adjustments to the plows in order to prevent damage to track-side structures as the train passed with its out-stretched plows.

For the removal of light snow, steel plows were mounted directly to the front of three of the four locomotives used by the Tanana Valley Railroad. The exception was the little locomotive No. 1, whose main job was shunting cars in Chena and Fairbanks. It was also used for the run between these two towns. While special track cleaning runs were made by these locomotives with their attached plows, it was not uncommon to see them also pulling, at the same time, a full train as it headed out to the creeks.

Little Eldorado Station stop. Notice the large snowplow and the link and pin coupler on the front of the locomotive.
Photo courtesy of the Candy Waugaman Collection

When the snow became too deep or compacted into the consistency of ice, the railroad would have to resort to brawn. Hundreds of laborers would be called out on the line, with picks and shovels in hand, to give a go at it. Since these drifts, quite often, towered above the snow shovelers, all the ingenuity of man was called into play — for it would take both man and machine, working symbiotically to finally clear the track.

It was during the bitterly cold winter of 1921 that the fabled hardness of the snow, which seemed to be confined to that stretch of track between Summit and Ridgetop, became unconditionally confirmed. Temperatures hovered at the - 60 degree mark for weeks, disrupting all transportation and community affairs. The Tanana Valley Railroad, which proudly extolled the fact that it was "an all weather transportation system," was forced to do some fancy foot-work to justify the shut down of its rail service. Taking a tongue-in-cheek approach, it attempted to rationalize this disruption of service by stating, "That during this cold weather, nobody wants to travel anyway, and therefore, there is no need to run trains over the track just for the object of running trains — meanwhile, they *(the trains)* will rest and keep warm!"

Fairbanks Daily News-Miner,
December 10, 1918

Winter's grip on the railroad was unremitting virtually every year. On January 2, 1921, the regular passenger train left Fairbanks, seemingly for Chatanika, as scheduled. Snow whipped up by strong winds left impregnable drifts along the track. In the process of trying to break through these massive mounds of snow, the locomotive broke its "permanently" attached snowplow, which had replaced the "cow-catcher" on the front of the engine, as it attempted to buck through these drifts.

The Fairbanks station in winter. The people shown all appear to be railroad employees busy at work. Notice the tracks are snowfree, while some of the community boardwalks are not. Notice the frost on the pole to the left of the picture.
Photo credit: University of Alaska Fairbanks, Alaska and Polar Regions Department Acc. #64-29-89, in the archives, of the Driscoll Photograph Collection

Totally defeated, the train slowly and very deliberately backed down the tracks to return to Fairbanks. Caution was the key word here; for if the snow ever got under the plow, during this reverse movement, it would lift the locomotive right off the track and cause a serious derailment. All the creek passengers aboard the train were forced to return to Fairbanks, mean spirited and disgruntled — but rationalized their delay by venturing back to their favorite "watering hole", to tank up with 70 proof "anti-freeze" in preparation for the next day's attempted rail assault on the hill.

With the repair of the plow on locomotive No. 52, the passenger train once again departed for Chatanika on the following day. It was anticipated by the crew of the train that this was going to be a very difficult trip, for there was great concern as to whether or not the frozen blanket of snow on the tracks would yield to the blade of the plow. Just beyond Summit, the engine, while taking a curve, hit a snowdrift with such a force that it lifted the locomotive just high enough off the track that it climbed up and over the rails. Deacon Jones, who was at the throttle for the first time on a creek run that winter, was unable to stop the engine. Initially everything seemed to be in order, until it was realized that the train was not, indeed, on the rails, but was running along the frozen snow making its own "tracks" as it went along. The rails had been left far behind some time ago!

"Live" locomotive pulling a "dead" locomotive.
Photo credit: University of Alaska Fairbanks, Alaska and Polar Regions Department Acc. #91-224-01, in the archives, of the Robert A. Lilly Photograph Collection

It was soon determined that the train could not make it back to the tracks by itself and that it would need the assistance of another locomotive. Conductor Rodebaug detrained and walked back along this forbidding terrain of a frozen white hell, to a section house at bridge No. 5 to telephone for help! Agent Raap at the Fairbanks depot received the message and called North Nenana to obtain help from the only other locomotive on the roster at that time. It raced up the track to Happy and switched over to the mainline for the dash up to Summit. Upon arrival, it coupled onto the last car of the stranded train and gently reversed its movement. The compacted snow was of such tenacity and strength that it did not crack under the weight of the de-railed train as it was pulled back to the original spot where it had left the rails.

Although the train was now back on the track, it could not proceed any farther. The line ahead was inundated with snow. What few passengers that were on board the train elected *not* to walk to Olnes, where food and warm shelter was available, but decided to spend the night in the passenger car. Food was brought to them by the train crew from the No. 5 section house. With stomachs full, a safe, warm, and comfortable night was spent in the coach. This derailment ultimately affected the whole operation of the line, for the only two locomotives operating at this time were tied up on the ridge. Thus, the run down to North Nenana was cancelled for the following day — and for several days to come. However, the mail did go through, for once again man reverted to the safest and most dependable travel in the north — the dog team!

Great concern was given, at this time, relative to the great volumes of freight which had arrived, but stagnated in Olnes. It was not only a mining region, in its own right, but also a distribution point for the Tolovana, Brooks and Livengood Creeks — and this freight was not being delivered out to the mining creeks where it was badly needed. Extensive flak was cast upon Mr. Landwehr, the agent in charge, by the miners for not getting out this freight to the camps. This occurred at a time when the railroad was struggling through a period of lean financial years and was quite sensitive about the image that this situation might create — at a time when the best possible service was utmost on the minds of the management. As a result, a strong admonition came down from Mr. Joslin himself, to the effect that every employee connected with the system would do everything possible to see that the public got its money's worth. In no time, the freight jam was broken and the freight was out on the trails to its respective destinations. No sooner did the railroad resolve this issue than it was struck by another "bomb shell" the following December. It seems that C.W. Joynt, the former general manager of the line, *brought suit against the company* because of failure to pay back wages amounting to $1,750. To compound this financial request, he was asking for an 8 % interest on this amount for delayed payment. Little did the railroad realize that this was only the prelude to many disasters that were to beset the operations and pocket books of this company.

Winter was now getting a grip on the land and the first environmental harbinger of what awaited the railroad during the coming December was on its way. On Thursday, December 2, 1915, the train, while on its way from Olnes to Chatanika, broke an axle on one of the trucks of a passenger coach. Since the

Paragraph 1 references Fairbanks Daily News-Miner, January 4 & 5, 1921

The Tanana Valley Railway

Wishes the People of Alaska, Especially Those of the Interior

A Merry Christmas and a Happy New Year

DURING THE COMING YEAR WE EXPECT THE GOVERNMENT RAILWAY TO START WORK HERE AND CONNECT WITH OUR LINE, FORMING A CONTINUOUS LINE FROM THE COAST TO CHATANIKA, WHICH FROM THERE WILL BE EXTENDED TO THE YUKON RIVER.

Fairbanks Daily Times, December 25, 1915
Little did the public realize how long it would take for the government railroad *(Alaska Railroad)* to actually arrive in Fairbanks (1923)

Fairbanks Daily Times, November 5, 1915

car could no longer proceed on its way, the passengers were transferred to a boxcar. One can only conjecture as to what the prevailing attitudes were among the passengers at this time; but, when one realizes that gallons of "hootch" were regularly brought into the creek towns in these boxcars from the two breweries in town, it is reasonable to assume that this could have been one of the happiest bunch of passengers to arrive in Chatanika in a long time!

Fairbanks Daily Times,
December 5, 1915

No sooner had this event passed, than the new year struck with a winter vengeance. It was January, 1916, when cold, gale winds and heavy snow struck the line at one of the worse possible places on the railroad. This was from Summit, milepost 28 to Ridgetop, milepost 29. What unfolded was a clear expression of human failure acting in contradiction to experienced advice and logic.

The train for Chatanika has stopped on an incline to replen ish its wood supply. Notice the fireman with a substantial log in his arms as he is about ready to load it onto the tender of the locomotive. In the background is trestle No. 5, 502 feet long.
Photo credit: University of Alaska Fairbanks, Alaska and Polar Regions Department Acc. #79-41-66N, in the archives, of the Falcon Joslin Photograph Collection

There was a long standing admonition to the engineers of the locomotives, to the effect, that should blowing snow be encountered on the north slope of the hill upon reaching Ridgetop and the crest at Summit, one was to proceed no further. An experienced crew would recognize these conditions and indeed, proceed no farther. On the other hand, if an inexperienced crew attempted to pass through this storm it could be catastrophic. Such was the case with engineer John Galloway. After receiving a briefing from agent G.E. Jennings in Fairbanks, the impatient engineer fired up his locomotive and chugged off, ignominiously, into relative oblivion for *two weeks*! He had assumed that what he saw blowing across the now buried tracks to be an insignificant accumulation. With one hand on the throttle and the other hand on the whistle cord, he rode off into the arctic storm. Coaxing all the power that he could muster out of the locomotive, he gradually advanced the throttle to the maximum degree only to find out that the train was gradually slowing down, and — then, most abruptly, the train came to a sudden standstill. The train had ridden into a wall of snow, seven feet deep! Quickly, and without let up, the fine flakes of swirling snow filled in the areas between the wheels, between the cars, and all the track behind the train. It was now firmly frozen to the track, and gradually was covered with layer upon layer of snow so that it blended in with the white expanse that reached off into the horizon.

Famous last words — Engineer John Galloway to station agent G.E. Jennings, *"Yes, Guy, I understand everything you have told me and am sure everything will be all right."*
Frontier Times, July 1969

Great concern overcame the people in Chatanika, for the long overdue train had not arrived, and nothing was known of its whereabouts. A dog team was sent out from Chatanika in search for the snowbound train. Upon finding it, the musher reported the situation to the management in Fairbanks, who quickly rounded up as many men as was possible to take their chances up at Ridgetop in an attempt to *dig* out the train. After *10 days* of relentless, back-breaking work, the train was finally freed from its icy mold. In the meantime, the crew and passengers survived by eating up the eggs and frozen bread that the train was hauling out to the miners. One has never heard anything further about the locomotive engineer!

Fairbanks Daily Times,
January 5 & 6, 1916

It was the beginning of the worst of times, for during the next three months it was one bad scenario after another. Continuously, during the month of January, 1916, the train was repeatedly delayed by snow storms — taking as long as eight hours, for a four hour trip, from Fairbanks to Chatanika. It was not uncommon for the engineer to stop his train while enroute to his destination, climb down from the engine with a shovel in hand, and proceed to manually clear the track of drifts — as he literally inched his train through the Ridgetop winter devastations. Even when it was not snowing, gale winds would inundate the track with high mounds of snow while at the same time compacting these flakes of snow into steel-like ridges resilient enough to overturn a train. Further, the whole region was covered by a winter rain, which was almost unheard of at this latitude. The whole line, from Gilmore to Chatanika, was turned into one gigantic hill of ice! Accompanying these rains were fierce winds that made it unsafe to walk about even in Chatanika.

For those who are familiar with the north country, one knows that when a cold snap hits, physical as well as emotional resiliency tends to become brittle and fragile. The brittleness of metal when exposed to extreme cold suddenly became quite evident during an incident when the train was on its way to Olnes from Ridgetop. As the train was descending the hill from Ridgetop, the flange on one of the tender wheels cracked and came careening off. Upon reporting the situation to the dispatcher in Fairbanks, a rescue train was sent out, in spite of the miserable and dangerous conditions, to bring a replacement wheel-axle unit to the derelict train. Under extreme climatic duress, the damaged wheel-set on the tender was replaced. Once replaced, the train continued on to Chatanika where, because of the increasingly severe conditions, it made an immediate turn-around and plowed its way back to Fairbanks. No passengers were accepted on the relief train; however, priority freight such as food was accepted but only as far as Olnes.

Since snowstorms could strike at *anytime of the year*, all road locomotives carried permanently attached snowplows on the front of the engine. This was usually sufficient to clear the track — but not always. On February 21, 1916, the railroad's make-shift snowplow was pressed into action. Snow had accumulated to immeasurable depths on the north ridge from Olnes to Ridgetop. Slowly backing up, the combination locomotive and snowplow distanced itself about 75 yards from the snowdrift. With smoke belching profusely from its stack, and with wheels spinning at such a rate that the unit was on the verge of getting out of control, it charged, and charged and — charged again into the snowdrift. But — to no avail.

The Gilmore crew had anticipated an early return of this snow train; but as the day was now coming to an end, a great anguish overtook them. Sensing trouble, they steamed up their locomotive and pressed up over the crest in search of the missing train. As they descended down from Ridgetop, they came across the moribund snowplow train, encased in snow and ice, but — still on the tracks! Ever so gently the rescuers eased their locomotive into the coupler of the tender of the stalled train. After two short

Paragraph 4 page 90 (which concludes on the top of this page) references Fairbanks Daily Times, February 12, 1916

Paragraph 1 references Fairbanks Daily Times, February 18 & 19, 1916

THE WEATHER — December, 1914.

*Hamilton, Mont.			**Fairbanks, Alaska.		
Date	Max.	Min.	Date	Max.	Min.
1	40	15	1	24	5
2	40	19	2	18	— 5
3	42	24	3	20	5
4	41	18	4	12	0
5	37	17	5	1	— 3
6	32	11	6	16	2
7	21	12	7	22	1
8	20	— 1	8	28	— 1
9	17	— 2	9	10	— 5
10	15	— 5	10	5	— 6
11	12	— 4	11	26	4
12	16	— 4	12	23	10
13	12	—12	13	17	5
14	7	—13	14	20	— 5
15	14	— 4	15	34	20
16	13	—10	16	29	7
17	12	—14	17	22	— 2
18	11	—13	18	24	14
19	12	— 9	19	32	18
20	19	— 7	20	33	6
21	27	— 6	21	18	— 3
22	41	12	22	19	12
23	44	20	23	12	— 6
24	41	13	24	12	— 6
25	—	—	25	1	—16
26	34	7	26	1	—20
27	38	23	27	12	—12
28	40	26	28	10	— 4
			29	14	— 2
			30	18	7
			31	18	—12
Average 15 above.			Averagc 10 above		

*Taken from "The Western News" Hamilton, Mont. Dec. 29-14.
**Taken from the "Red Cross Drug Store" weather statistics

Citizens with ulterior motives of attracting yet more settlers to Alaska often downplayed the severity of Fairbanks winters. Judge Wickersham published the following excerpt in the *Alaska Central Railway Bulletin* in 1906.

"A magnificent bracing climate, no colder than the climate of the northern states, and it is extremely healthful..."

Likewise, it was usual to see charts favorably comparing Alaska's weather with that of various northern states in the early newspapers. The above chart was published in the Fairbanks Daily News-Miner, January 7, 1915.

In truth, it can be said that Seward has a climate similar to Boston, Anchorage is similar to St. Paul, but Fairbanks can be compared to no other!

Paragraph 1 references Fairbanks Daily Times, February 23, 1916

TRAINS TIED UP BY SNOWSTORM

EASTERN TRAFFIC IS NOT MOVING—NEVER HAPPENS IN THIS LAND.

POUGHKEEPSIE, N.Y., Dec. 14.—Three are dead, two are missing, and hundreds of passengers are imprisoned in twenty-five trains which are tied up near this city unable to move because of the fierce blizzard that is raging.

Taft Is Snowbound.

NEW YORK, Dec. 14.—Among those who are on one of the New York, New Haven & Hartford trains that is snowbound is William Howard Taft, who was en route to New Haven from this city. Many of the seaboard lines are badly crippled and traffic is at a standstill on all roads out of New York.

Freight Tied Up.

ALBANY, N.Y., Dec. 14.—There are no trains moving in either direction today. Freight trainis are tied up, and passenger trains that start out on their scedules have not been reported, and are blocked somewhere by the storms.

WARM WEATHER IN INTERIOR

Bulletin Shows Thermometer Ranging From 31 to 62 Above Zero.

According to the signal corps bulletin of this morning the sun is shining allover Interior Alaska and it is clear and calm. Everywhere the snow is disappearing rapidly and indications are that the water from the hills will be pouring into the larger streams within the next day or two.

Another mail left Copper Center for Fairbanks this morning. The stage also had some passengers.

The signal corps bulletin is as follows:

Salcha, 50; clear, calm, snow gone on trail.

Richardson, 40, McCarty, 40, Donnelly, 62; trail fair.

McCallum, 35; trail soft.

Paxson's, 39, trail going fast, 27 inches snow on level.

Hogan's, 31, trail bare.

Gulkans, 33; light wind, trail good.

Copper Center, 46; trail good. North mall left with first-class States mail at 6 a.m.

Tonsins, 45; trail fair.

Teikhell, 44; 4 inches snow, trail bare

Beaver Dam, 40, Valdez, 48.

Nenana, 45. Minto, 45. Tolovana, 46. Eureks, 55. Rampart, 49. Hot Springs, 50. Gibbon, 40. Nulato, 41.

Clear and calm all stations except as noted. Find day over system.

Fairbanks Daily News-Miner, December 14, 1915

blasts from *both* engines, their drive wheels very deliberately gripped the rails as they slowly rotated, trying desperately to avoid spinning out. As luck would have it, the train gradually picked up momentum with each revolution of the drive wheels. The wall of snow and ice ahead of the snowplow suddenly began to crack — and, with a thunderous roar, all of its stored up energy let loose, as it was cast aside like an ocean wave ready to strike the shore. With whistles screaming like a banshee in the middle of the night, the two locomotives kept up their charge into the snowbank, without missing a revolution of the wheels — continuing on across the flats, smashing one snow drift after another all the way to Chatanika!

These drifts were six feet deep, and the snow was crusted so hard that it had the consistency of stone. While there were many minor hardships during this winter's ordeal, none was more embarrassing than the one that befell one of Chatanika's finest, Erick Heggebloom. During all this winter turmoil, he had passed away. His funeral was to be held in Fairbanks, but with all the disruption of railroad operations because of the snow, wind and cold, he was relegated to the status of non-priority freight. Consequently, he was to suffer from the ignominy of being late for his own last rites and funeral.

March had come in like a lamb. Drips of water in ever increasing amounts were beginning to fall from the trees. The sun was getting much higher and it was the feeling among the locals that the "Hawk" would soon be summarily dispatched, for the pussy willows were peeking out from their hibernation. Yet, in spite of these early signs of spring, this was not to be.

A series of winter storms was about to unleash itself upon the region, the likes of which had never been experienced before. From Fox, milepost 18, the line was completely erased from view by the heavy cover of snow. The regularly scheduled train was able to go only as far as Goldstream (a mining enclave), which was somewhat before Fox. Unable to proceed any farther, it elected to *back* its way to Fairbanks. In the short time that it took to make the decision, the whole right-of-way to the rear of the train had become covered by highly compacted snow. While there was some thought of sending out a rescue train, there was that fear that all of the locomotives of the railroad might wind up stalled in the snow drifts, indiscriminately, along the whole extent of the line. Management elected, for the time being, to shut down the whole railroad, something that it did not do during any period in the history of the railroad.

This unmitigating arctic blast continued on well into the third week of March, 1916. Between Gilmore and Olnes the track and the railroad structures had seemingly disappeared. The blowing snow was so fierce that it had reshaped the contour of the land by creating high mounds and drifts of snow all over the terrain. Fifty men with shovels, and two train crews were pressed into service in an attempt to open up the line. The two train crews, however, had to share the only locomotive available between themselves, for the other three were marooned somewhere out on the line. To further compound matters, the spring-like weather had caused alternating thawing and freezing every night. As a result, everything was now in the firm grasp of the newly formed ice.

Some semblance of operation was attempted during this period when the storm closed the line. Passengers and freight were finally accepted, after the initial delay, but only with the stipulation that the railroad would transport them *only as far as the train could go*. In most cases, this turned out to be only as far as Gilmore. Mail service, however, was only partially affected, for the railroad dispatched the mail out to the creeks via dog teams. When the line was finally opened a week later, the railroad, out of respect for the uncertainty of the weather, would transport passengers only as far as Gilmore.

Arriving in Gilmore, the passenger car was uncoupled from the rest of the train and was placed on a siding. The locomotive would then *recouple onto the freight cars*. With weather permitting, it would challenge the hill and, with luck, bridge the crest to continue on to Ridgetop and Chatanika. This precaution was taken because the snow had accumulated to such great heights that the train would have to actually pass through a moat with mounds of snow on either side. The chance of an avalanche occurring as the train passed through it was an ever present danger. Shortly after March 25, the railroad resumed its "normal" operations, never wanting to look back to the winter of 1916.

Once winter's storms had abated, the coming of spring did nothing to alleviate the railroad's operational problems. Spring breakup of winter's snows caused extensive flooding of the line, especially along the whole of the Goldstream Flats from milepost 9 to milepost 16. In May, 1911, the area became one grand expanse of water. May 7 of that same year was particularly memorable because the water was so deep that it put out the fires in the boilers of the locomotives as they attempted to pass through this "inland sea!" Waves were rolling over the water to such a degree that they created a feeling among the passengers that they were at sea. This constant motion caused both horizontal and vertical movements of the passenger coaches — an affect that caused many of the passengers to be on the verge of becoming *seasick*! However, the greatest difficulty was between Big Eldorado and Engineer Creek where train movement became so unsteady that it was analogous to a maritime passage. Facetiously, it was proposed that screw-propellers or paddles be attached to the locomotives when the thaw came, for then "they could be operated by a licensed pilot!"

As the water drained off the surface of the land and into the streams and rivers, it traveled at such a velocity that the trapped ice flows would be driven into the railroad bridges with such intensity that extensive damage was done to the pilings. Almost annually, the most victimized bridge was the trestle over *Dead Man's Slough*, not far from Fairbanks. To insure that the passengers would be able to reach their destinations during the periods when the bridge was considered unsafe for train crossings, yard engine No. 1, the little 0-4-0 was pressed into a unique service.

SIGNAL CORPS STORM REPORT

The chief operator of the telegraph system tells a good story about his experiences last night, when he started to make inquiry as to the condition of the weather. First he called up Salcha and in answer to his question, the operator said, the weather was "pretty bad." Next he called Richardson, and the reply was "fierce." McCarty was then called and the answer was, "It's h—l." He was afraid to call the next station and quit.

Fairbanks Daily News-Miner headline, January 5, 1916

Paragraph 2 references Fairbanks Daily Times, March 17 & 25, 1911

CREEK PEOPLE LEAVE FOR HOME

Many Who Were Held On Account of the Cold Weather Leave On Train.

Fairbanks Daily News-Miner headline, January 25, 1916

Paragraph 3 references Fairbanks Daily Times, May 8, 1911 and Fairbanks Daily Times, April 6, 1909

SNOW AND COLD STOP SLUICING

Indications Are That Break-up Will Not Be An Early One

Fairbanks Daily News-Miner headline, April 21, 1916

Our Railway Going Strong

The T.V.R. yards and shops on Garden Island are taking on a lively air these days.

The new steamshovel, which has been sent out to bridge 5, where it will be used on the big fill there. Engine 50 has been put in commission and, with J.H. (Deacon) Jones who ran the dinky last summer at the throttle, will be used in connection with the steam shovel the coming summer.

A new track piledriver has also been built in the local shops, for putting in any of the trestles which may be carried out by the ice at the breakup, the bridge across the big slough having been carried out that way sevveral times in the past winter, there will likely be high water at the breakup.

Fairbanks Daily News-Miner, March 20, 1920

Paragraph 2 references Frontier Times, July 1969

Paragraph 3 references Fairbanks Daily News-Miner, April 24, 1919

Fairbanks Daily News Times, August 17, 1906

Just prior to the onslaught of the winter's ice-flows, the locomotive was brought up from Chena with a passenger car and stationed on the Fairbanks side of the slough. As the train from the mining creeks approached the bridge, it gradually slowed down so as to get as close to the bridge as possible. Passengers would then disembark from their coach and proceed by foot to the other side of the trestle. Once across, they would board the waiting coach with old No. 1 on the front end, for their final passage into town.

After the deep waters of spring receded, the long summer days would cause a prolific growth of high wild grasses. In spite of all the efforts of the section gang to keep it trimmed, the wild grasses continue to proliferate to such degree that it would ultimately fall over onto the track. This would make the rails so slippery that the trains could not develop the traction to proceed. In the autumn, these same grasses became a source of repeated fires, for by then they had become thoroughly dried out, and therefore, were in a potentially combustible state. As the train passed through this sea of grass, it would become easily ignited by any stray spark from the locomotive. Not only was the grass ignited to cause a serious fire, but also the ties, and any other bridge or wooden culvert that was traversed by the train!

On one occasion, a Vault resident accused the railroad of deliberately starting a fire which had caused him, and others, great harm. He claimed that the railroad sent out its own people to ignite the weeds and grasses on the right-of-way to "protect their *dinky* tracks!" During one of these burns, the fires became so uncontrollable that they traveled downhill to engulf the timber on the lower slopes. This caused a raging inferno which ultimately burned out some of the miner's cabins and woodpiles.

Not only did the tall grasses grow prolifically during the summer months, but so did all the vegetation. Particularly abundant was the annual crop of blueberries, which ultimately became the main ingredient of an elegant wine with the consistency of a California Tokey. This wine was made by the miners themselves to be served at dinner and to warm the heart and soul during those long, lonesome, cold winter nights. On other occasions, however, the miners loneliness and isolation could, with the help of this blueberry wine, lead one to acts of desperation. Such was the case of Patty McGuire of Engineer Creek. He decided that with the help of the Tanana Valley Railroad he would call it quits and leave the struggles of life behind. On May 26, 1913, he stretched himself out across the rails at a point between Fox and Engineer "cities" where the track made a sharp curve shortly after leaving the former community. As the train rounded the blind curve at a point where it was impossible to see the track, the engineer was just able to make out that there was an "object" across the rails. He jammed on the brakes and waited for the inevitable crash as the momentum of the train kept it moving forward. Literally, within a matter of a few feet, the train finally came to a complete stop. The brakeman and the conductor of the train ran forward to see what was across the track. Much to their surprise there was Patty McGuire, lying across the tracks in a state of intoxication — but still alert. His only reaction to this situation, upon being removed from the track, was a strenuous objection. He felt that is was none of the business of his "rescuers" if he wished to die. Since the next train over that piece of track wasn't until the next morning, one must assume that Patty had plenty of time to think it over after he became sober!

Fairbanks Daily News-Miner, May 26, 1913

While winter's icy talons raised havoc with the railroad's operations, its affect on shipping was equally insidious, dramatic and final. Indeed, freeze-up in the autumn is an expected phenomenon in the north country, but its exact day of arrival, on the other hand, is not. Riverboats, at this time of the year, began to take chances in order to complete their last run of the season, not-with-standing the fact that they might become ice-bound. When the temperature drops below 32 degrees (F), crystals begin to form on the water, and the flow of a river becomes sluggish. Gradually the ice crystals fuse to form larger and larger ice masses, which subsequently begin to plague river traffic as the slush impedes forward progress. Suddenly, traffic comes to a halt and it is the beginning of the end for all shipping until the following spring.

ANCHORAGE IS NOW ICEBOUND

(Special to News-Miner)

SEWARD, Nov. 23 — The Alliance has returned to Seward with a full cargo from Anchorage. The boat was unable to unload there owing to the great amount of ice. The ice prevents the landing of the boats. The Northwestern arrived here last night with four carlolads of rails, which the captain brought to Seward rather than risk losing them while unloading at Anchorage. There is a bargeload of rails adrift in the Inlet now.

Only four of the nine barges whch the commission bought are left. The others have been lost. The wreckage is still on the beach. The Kansas City is at Port Graham, where the vessel is undergoing some slight repairis. The Admiral Farragut left Seward for Anchorage, but will probably not make it.

The freighter Seward sailed last night with a big lolad of freight for Anchorage, but it is feared that the vessel will not be able to make it and will have to unload in Seward.

Fairbanks Daily New-Miner, Novermber 23, 1915

On Wednesday, October 6, 1909, such an incident occurred, *en masse*, when a multitude of riverboats found themselves suddenly locked in ice for hundreds of miles up and down the Tanana and Yukon Rivers. While some of the boats were able to make it to shore or a convenient slough to sit out the winter in the wilderness of Alaska, most came to an inglorious stop right in the middle of the river. At times, these ice bound steamers were in such a precarious and awkward position that the following riverboats could not pass by and were themselves doomed to be engulfed by the surrounding ice mass.

During the winter of 1909, the riverboats *Herman, Light and Hamilton* attempted to beat out the developing ice by making a dash, full-steam-ahead, for the sea at the mouth of the Yukon River. Within view of the open ocean, they suddenly became locked into this icy vise and were not able to break through this relatively short distancc to freedom.

Paragraph 3 references Fairbanks Daily News-Miner, October 6, 1909

Farther upriver, another drama was taking place. The *Koyukuk* with passengers and freight from Dawson in the Yukon, just made it into the mouth of the Chena River, narrowly escaping from the massive ice flows which were beginning to hammer at the hull of the boat. *Seattle No. 3*, and the *Schwatka*, which were following the *Koyukuk*, were not so fortunate and were forced, very abruptly, to go into winter quarters in the bush at Oayne's Cutoff, 25 miles above the Tolovana River and 150 miles from Fairbanks.

Fairbanks Daily News-Miner, October 6, 1909

Crews from the stranded vessels, most of whom were residents of other states, would cross the ice by foot from their ice-bound vessels. They then would make their way, as best as they could, by foot to

Fairbanks and then by sleigh to Valdez. This was a 364 mile trek through the rugged Alaska and Chugach Ranges, taking nine days to complete the trip. Normally, the crews would have tied up their boats for the winter in St. Michael and then board an ocean-going vessel for Seattle.

Author's Note:
Five steamboats were built at Chena and Fairbanks between the years of 1905-1913. They were:
White Seal - built 1905, weighing 194 tons
Teddy H. - built 1910, weighing 153 tons.
Samson - built 1910, weighing 272 tons.
Idler - built 1911, weighing 61 tons.
Shusana - built 1913, weighing 49 tons.

The *White Seal* was purchased in 1905 by the Tanana Mines Railway. The record accounts are a little hazy, but *White Seal* apparently was hulled on a rock and partially sunk, shortly after purchased. She was raised and repaired at the Chena docking yards. But repair costs caused the railway to balk and ownership reverted to the original builders.

With so much freight stranded along the river channels, many measures were taken to retrieve as much of it as possible for immediate use. Usually, however, the steamship companies *would not* bring into town any of the stranded freight. The consignee was forced to seek out the assistance of another service, such as the Pinkerton or Fairbanks Transfer Companies, to haul by horse-drawn wagon, this freight across the wind swept ice into town.

The entire cargo from the *Schwatka* was removed from the boat and stored, first, in a crudely built log warehouse in old Minto before being transported into town. Since Minto was a good 55 miles from Fairbanks, one had to wait for a heavy snowfall, so as to improve the conditions on the trail for the sleighs, before moving the cargo. After being reassured that the perishables, including some very important supplies for the *Arctic and Barthels Breweries*, were still in good condition, Messrs. Volney Richmond and Reed Jones of the Northern Commercial Co. gave the order to bring this freight into town. Three teams a day were used to bring in 500 tons from the *Schwatka* — the last team of the day always dragging some spruce boughs on the rear of the sleigh. This would brush the snow into the ruts caused by the frequent passage, which reduced the roughness of the trail. Unfortunately, because of the extreme volume of the cargo and the limits of the horse-drawn sleighs, the greater portion of the goods had to remain behind until spring in the specially-built log cabin.

In another similar scenario, the *Julia B.* had the cargo removed from the vessel and stored for the winter in a log cabin constructed just below Twelve Mile Bar. A watchman was dispatched to the scene whose primary duty was to keep a fire going in the cabin, all winter long, so as to prevent the cargo from freezing up!

Paragraphs 2 & 3 references Fairbanks Daily News-Miner, October 2, 1909

The ultimate alternative to stalemate on a frozen river was to proceed in haste to Chena, hopeful to be able to spend the winter in town. Since many boats would arrive almost simultaneously during the freeze-up, an enormous amount of freight was left on the dock at one time. Consequently, freeze-up put a formidable burden on the transportation abilities of the Tanana Valley Railroad. Under normal conditions, autumn was an extremely busy time for the railroad in its own right. When this added freight was suddenly placed on its rails, it appeared to be only a matter of time before its operations would grind to a halt. Train after train, once loaded, would immediately depart for the mining creeks, for these supplies were badly needed at this time, to prepare for the extensive mining operations which were to occur the following spring. These frequent train movements would continue on until the new year, when at last there would be a reprieve from this hectic schedule, and the railroad could now return to its usual operating schedule — a single, daily round-trip to Chatanika!

Fairbanks Daily News-Miner,
October 23, 1923

TANANA MINES RAILROAD
PIONEER RAILROAD OF
INTERIOR ALASKA
Four Times daily between
CHENA — FAIRBANKS —
ALL MINING DISTRICTS
Falcon Joslin, President &
Manager
C. Morarity, Superintendent
J.H. Scott, General
Manager

Tanana Mines Railway advertisement, Fairbanks Daily News-Miner printed October 29, 1906, showing 4 trips per day

Tanana Valley R. R.
TIME CARD.

To Gilmore, daily except Sunday.
To Chatanika Mondays, Wednesdays and Fridays, only.

**IN EFFECTWEDNESDAY,
SEPTEMBER 3, 1913
DAILY EXCEPT SUNDAY.**

Leave Fairbanks	9:00 a.m.
Leave Ester Siding	9:25 a.m.
Leave McNeer	10:00 a.m.
Leave Fox	10:45 a.m.
Leave Gilmore	11:15 a.m.
Leave Ridgetop	12:10 p.m.
Leave Olnes	12:40 p.m.
Arrive Chatanika	1:05 p.m.
RETURNING	
Leave Chatanika	1:45 p.m.
Leave Olnes	2:10 p.m.
Leave Ridgetop	2:45 p.m.
Leave Gilmore	3:40 p.m.
Leave Fox	3:55 p.m.
Leave McNeer	4:20 p.m.
Leave Ester Siding	4:40 p.m.
Arrive Fairbanks	5:00 p.m.
FOR CHENA — Daily Except Sunday.	
Leave Fairbanks	5:25 p.m.
Arrive Chena	5:55 p.m.
RETURNING	
Leave Chena	7.15 a.m.
Arrive Fairbanks	7:45 a.m.

Parker's Auto connects at Ester Siding.

Eagan & Griffin's stage connects with auto car at Gilmore Mondays, Wednesdays and Fridays. Stages connect at Gilmore and Chatanika.

**C. W. JOYNT
General Manager.**

Fairbanks Daily News-Miner, 1914 time table showing winter schedule or one run perday

CLEMONS PHOTO FAIRBANKS, ALASKA. 1918.

CHAPTER 10

THE LONG TWILIGHT

Mogul No. 51 at a fuel stop in Fairbanks. Wooden sides were added to the tender in order to increase the volume of fuel it could carry.
Photo courtesy of the Candy Waugaman Collection

Alternative train travel on the Tanana Valley Railroad. Location is unknown.
Photo courtesy of the Colin MacDonald Collection

Within a period of five years, that is, from 1905 to 1910, the railroad reached the period of its greatest productivity and financial gain. Then, almost abruptly, it entered into a period of gradual decline that lasted for seven years. Ultimately this decline led to its demise in 1917 as an independent operating entity. While the genesis of this predicament was a result of many factors, it was an act of federal intervention that rendered the death knoll, indirectly, for the financial devastation of the railroad.

In 1906, and again in 1909, President Theodore Roosevelt, by edict, withdrew from public use all Alaska coal, oil and forest reserves. This act effectively blocked all industrial development in the territory. Interestingly, the edict was not rendered by the president of his own volition, but rather was forced upon him by political pressure that was instigated by coal and timber barons in the states who feared competition from Alaska. As a result, a badly needed source of energy was choked off from the railroad, the mines and the Interior communities.

The primary source of energy, up to this time, was wood. With the closure of the forest reserves, a hold was placed on every aspect of life and commerce in the region. The voracious appetite of the developing communities and industry for energy had virtually depleted the available land of its trees. This lack of wood placed a heavy financial burden on the Tanana Valley Railroad for it was fast becoming the most expensive service item on its purchasing agenda. Initially, the railroad had to pay $9.00 for a cord of wood that was stacked along its roadbed, and much more as this source of energy dwindled farther away from the right-of-way. Sadly for both the railroad and the mines, along with all the surrounding communities, coal, a primary source of energy, could have been tapped into just 60 miles south of this region if the federal government had not intervened!

To further compound the strangle hold placed on the Tanana Valley Railroad by the federal government, the Alaska Road Commission set about the task of completing a system of roads in and around Fairbanks and out to the creeks. Since these new roads, in many cases, provided a shorter distance to the mines, passenger and freight loads began to shift from the rails to the roads. For example, from Fairbanks to Fox it was 11 miles by automobile, while on the other hand, it was 18 miles by train. True to the entrepreneurial spirit of man, David Courtemanche, a Fairbanks resident, soon established a competitive automobile service to Chatanika. His "limousine" departed daily, at 9:00 A.M., from the old Nordale Hotel on Second Avenue.

Though these new roads were usable in the summer, they often were disastrous, generally, for automobile travel during the periods of break-up and freeze-up. Consequently, some traffic would return to the rails in the winter. Never-the-less, management felt that it was nothing more than a losing proposition for its gluttonous wood burning locomotives to haul a single passenger car with only three or four passengers on board, going, at best, only a short distance!

Tanana Valley R. R.

TIME CARD

EFFECTIVE FRIDAY, MAY 1, 1914

Subject to Change Without Notice

Train Daily / Except Sunday.

Leave Fairbanks	9:00 a.m.
Arrive Chatanika	1:05 p.m.
Leave Fairbanks	5:25 p.m.
Arrive Chena	5:55 p.m.
RETURNING	
Leave Chatanika	1:45 p.m.
Arrive Fairbanks	5:00 p.m.
Leave Chena	7:15 a.m.
Arrive Fairbanks	7:45 a.m.
ELECTRIC CAR DAILY	
Leave Fairbanks	4:30 p.m.
Arrive Gilmore	5:40 p.m.
RETURNING	
Leave Gilmore	8:20 a.m.
Arrive Fairbanks	9:30 a.m.

Automobile connects with Electric Car at Gilmore for all points on Pedro creek; Leaves Douglas (Summit) roadhouse at 7:30 a.m.; returning, arrives Summit 6:40 p.m.

Automobile connection at Ester Siding for Ester City.

Stage connection at Gilmore for Fairbanks Creek.

Stage connection at Chatanika for Cleary City.

Special trips with auto car made at any time with party of ten or more to any point between Fairbanks, Gilmore or Chena.

C. W. JOYNT,
Genl. Mangr.

Fairbanks Daily News-Miner. Railroad time card printed May 1, 1914. Note the automobile and freight connections listed at the bottom.

Paragraph 4 references Fairbanks Daily Times, November 7, 1909

Paragraph 1 references Fairbanks Daily News-Miner, September 2, 1909 and Tanana Valley Railroad Annual Report, 1909

To compete, the railroad cut its own passenger and freight rates to one-half of that charged by other railroads for a similar service. On Friday, September 3, 1909, reduced fares on the Tanana Valley Railroad went into effect. The fare to Fairbanks from all stations up to and including Fox City dropped from $2.00 to $1.00 one way, and $1.50 for the round-trip. Similarly, the fare to and from Gilmore dropped from $3.00 one way to $1.50 and to $2.00 round-trip; while the fare to Olnes, Little Eldorado and Chatanika was reduced from $5.00 one way to $2.50, and to $4.00 for a round-trip. In addition, 50 pounds of baggage would be carried free for each passenger. The railroad then invested in its own *non-rail* transportation network which radiated out from one to five miles into the creeks from certain railroad stations.

The Edison-Beach Storage Battery Car at the Fairbanks Railroad Station. The original station is pictured on the left, and the "new" station is on the right. In the background is the Chena River Bridge, which was later moved to Nome.
Photo Credit: University of Alaska Fairbanks, Alaska and Polar Regions Department. Acc. #68-69-896N, in the archives, of the Lulu Fairbanks Photograph Collection

While great expectations had been anticipated, the opposite prevailed. Although passenger loadings temporarily increased, the earnings did not. This was due to the fact that the travelers utilized the railroad only for short haul distances, and at a markedly reduced rate. Ultimately, it was clearly evident that regular passenger service would become a losing financial proposition. Falcon Joslin, therefore, ordered in 1909, for the relatively few passengers that continued to ride the train, an *Edison-Beach Storage Battery Car* as an alternative and inexpensive means of passenger rail transportation. This type of car had been used on some of the eastern railroads with reportedly good results. It was felt that this new addition would not only improve the service to Fairbanks and the creeks, but also enhance the reliability of rail service.

Since the Tanana Valley Railroad was a relatively short line, this battery powered unit served the company well, for the car had a range of 100 miles before recharging was again necessary. It was planned to use the car on three round-trips to Gilmore a day, with an additional round-trip to Ester Siding. A railroad operated automobile would meet the electric car at this siding to transport the Ester bound passengers to that destination. Additional parallel automobile service out to the creeks was also instituted.

Paragraph 2 references Tanana Valley Railroad Annual Report, 1910

Paragraph 3 references Fairbanks Daily Times, February 7, 1912

Paragraph 4 references Tanana Valley Railroad Annual Report, 1913

The use of this electric car was, in principle, an attempt to deflect those losses which had resulted from the increasing competition offered by the automobile. To some degree, it was able to offset some of these losses by carrying more people. However, the railcar proved to be ineffective in increasing revenues, for the trips involved were for only short distances. Even with a further reduction in passenger fares in 1913, the loss of passenger travel to the automobile could not be stopped.

As best as can be ascertained, the electric railcar ran primarily to Fox and Gilmore. On one of these runs to Gilmore on July 2, 1914, the Edison-Beach Storage Battery Car left the rails shortly after leaving Fox — falling quite a distance down the hill and coming to rest on its back with the wheels spinning wildly in the air. It remained in this position for over a week before attempts were made to right it and put it back on the tracks. Once it was placed in an upright position and back on the tracks, a most remarkable aspect to this accident came to light; although it fell a great distance, it suffered very little damage and was able to return to Fairbanks under its own power!

The car shown is gasoline-powered and the picture caption says "Motor car in use between Nenana and Fairbanks." No date given.
Photo credit: Anchorage Museum of History and Art. Acc. #AEC G 2028, Alaska Railroad Collection, H.G. Kaiser, Photographer

Schedules to these communities varied in accordance with the needs of the public, and those too, of the railroad. It made runs, also, to Chena, but interestingly, the runs were associated with periods of low water in the Chena River, which prevented the riverboats from coming upstream to Fairbanks. Accordingly, passengers departing Fairbanks for trips downriver to St. Michael were accommodated by the railroad by scheduling a special run to Chena, with the electric railcar, so that they could make their riverboat connection.

Paradoxically, on other occasions, the riverboats, during periods of low water on the Chena River, *could get up to Fairbanks*, but the passengers were not allowed to get on-board. When the riverboat *Reliance* left Fairbanks on midnight of May 20, 1916, it was forced to instruct the on-coming passengers to board in Chena, because the water levels were so low. Although the water level of the river might accommodate the draft of the riverboat with its loaded freight, the added weight of the passengers might just make the difference in causing the boat to scrape bottom and get hung up on a sandbar. The passengers were, therefore, forced to charter the electric railcar for their trek down to Chena where the riverboat would make an unscheduled stop for them.

Fairbanks Daily Times, May 20, 1916

Not only did the electric railcar find itself performing unique services on demand, but so too did the motorman who operated the unit. For a period in 1914, the electric railcar was run on a daily basis from Fairbanks to Gilmore and back. Since the miners were scattered all over the hills, it was not sufficient enough for the motorman to just drop them off at the depot in Gilmore. Parked at the station was a company automobile which was at the disposal of the miners. The motorman, upon arrival, would secure his "train", dash over to the automobile, crank up the unit and taxi his fares to their respective destinations, all over the hills. Upon completion of this highway run, he reversed his travels and returned to his "train" and electrically "hummed" his way back to Fairbanks.

Fairbanks Daily Times, April 30, 1914

By 1922 the electric car was out of service. However, at this very same time concern had developed as to what one might do to implement rail service between Fairbanks and College, the home of the present day University of Alaska. Trains whose ultimate destinations were Chatanika, or Seward could no longer meet the frequency needs of the residents of the area. President Bunnell of the college, and Col. Mears of the Alaskan Engineering Commission, appropriated the old Tanana Valley Railroad electric car and sent it down to Anchorage to be rebuilt from a straight electric car to a *gasoline-motor electric car*. Upon its return from Anchorage, it was christened the *Tanana*. It did not have the conventional four wheel trucks which are located to the front and back of the car, but instead, had only two wheels on the front, and two wheels at the back of the car, as is customary on the European railroad cars. Each wheel was driven by a 5 H.P. motor, giving it a total of 20 H.P. These electric motors received their energy from a 125 volt generator which was powered by a gasoline engine. The car speed was controlled with a rheostat similar to that used on the old time trolley cars. The gasoline motor had a governor on it, so that when it was running the engine man could set it on high idle to make and receive more electricity. Simultaneously, he could adjust the electric comptroller as needed. Except for the fact that it had a gasoline motor instead of a diesel motor, its principle for electrical production was similar to the present day diesel-electric locomotives.

1994 conversation between the Author, Nicholas Deely, and M. Maddox, former Tanana Valley Railroad/Alaska Railroad employee.

Known affectionately as the "Toonerville Trolley", it had the appearance of a small streetcar, but lacked the usual overhead trolley-pole. Although the Tanana Valley Railroad was a narrow gauge line, the body of this car was *standard gauge in width*. In effect, it gave the appearance, when in motion, of that of a waddling duck. It sported a headlight mounted both on the front and back on the unit, plus an additional large headlight mounted on the front end. Except for the cost of the motorman and conductor, operating expenses were minimal, averaging five cents a mile.

December 7, 1931 letter written by W. L. Kinsell, Superintendent, Motive Power and Equipment, Alaska Railroad to Mr. Metzdorf.

One of the motormen in charge of this unit was a chap by the name of Marsh. He got to know the students and passengers so well that if anyone was late, he was known to hold up the "trolley" pending the arrival of his anticipated passenger(s). His schedule consisted of daily departures from Fairbanks to College, except for Saturdays and Sundays, at 8:30 A.M., 12:15 P.M. and 4:00 P.M. Returns from College were at 8:45 A.M., 12:30 P.M., and 4:15 P.M. On Saturday, only the morning and noon runs were made. No trips were made on Sunday. The fare was 25 cents for the 15 minute ride. Having established this new, service oriented schedule, both President Bunnell and Col. Mears felt that it would be unconscionable to drop off the students along side a snow covered path which lead up to the college. Consequently, a structure was erected at this site which was to become the College Station on the forth-coming Alaska Railroad.

Paragraph 2 consolidates references from Fairbanks News-Miner, October 20th & the 30th, 1922

Through the years, the "Toonerville Trolley" provided efficient, reliable and safe service. However, on March 23, 1929 the gas-electric car, *Tanana*, abided by Murphy's Law; that is, if anything can go wrong, it will! Coming into the Fairbanks depot, the *Tanana* suddenly lurched to the side and tipped over. The passenger load consisted of eight people, three of whom were painfully injured by the flying glass. Of these, one suffered from an extensive laceration of the ear which required five stitches to close. One of the others was cut about the hands, but required no surgical care. Two other passengers were thrown out of the windows. The rest, after the car stopped rolling, were able to get out of the rear of the car with no difficulty. Oddly, the "*Tanana*" suffered from only a small amount of structural damage involving the brake rigging, broken windows with shattered glass, and jammed doors. After being placed upright on the tracks, it was shunted into the roundhouse for repairs. Since, at this time,

Electric car leaving Fox, June 29, 1915, with motorman Hanson at the controls.
Photo courtesy of the Candy Waugaman Collection

the track out to College and beyond to Happy was of dual gauge, a standard gauge car was substituted for this run until all repairs were completed on the "*Tanana*".

Because of the *Tanana's* oversized proportions, there was some thought upon the abandonment of the Tanana Valley Railroad, to replace its axles and wheels with standard gauge ones allowing it to maintain the Fairbanks-College run, but now on standard gauge tracks. This never materialized, for its place was taken by a *Brill Car*, which in affect was a bigger version of the same car. However, shortly after the onset of this service, it met with great disfavor. On one hand, it consumed too much gasoline and, on the other hand, a far more serious problem existed — its wheels were made of cast iron which had a tendency to *crack* and *fragment* easily!

In addition to the Electric Car and the Gasoline-Electric Car, the Tanana Valley Railroad possessed four *straight gasoline motor cars*. No. 5 was a 50 H.P. unit having a "Buda" motor in it, with a capacity of 17 passengers. Another unit had a 4 cylinder, Fairbanks-Morse 14 H.P. motor also having a seating capacity of 17 riders. Two other units were on the roster — both much smaller. The larger of the remaining two had a 2 cylinder, 14 H.P. motor with a seating capacity of seven. The last unit, had only a seating capacity of three people. It was powered by a 1 cylinder, Fairbanks-Morse motor which was air-cooled.

As early as 1908, during a down turn in business, the railway started to use these units instead of the regular train for its run out to Fox. It advertised this as a twice daily service, indicating that it would "make all the stops, between *here and that place*, to put off and to take on passengers!" Already many miners had broken camp out of financial desperation and were trekking out to the new gold fields in Iditirod. These motor railway cars accordingly appeared to be more than adequate to handle the resulting meager traffic.

As the years went by, the Tanana Valley Railroad came to rely heavily on these units to keep costs down. This imposed frugality led to some of its own experimentation. Fred Lewis, a Garden Island mechanic, constructed for this line a new motor car using a high speed gasoline engine. It was larger than the other cars — seating 20 passengers. It was driven by a 20 H.P., high speed engine, which was taken out of a motor truck that had been used locally for several years before its transformation from a highway vehicle. This unit was used on the Fairbanks - Chena run and also on a twice daily schedule to Gilmore.

These motor cars, or as they were then known, "speeders", served not only as a means of local transportation, but were also pressed into service as emergency vehicles at times of disaster and distress. This

Fairbanks Daily News-Miner, March 23, 1929

Freight train arriving in Fairbanks with locomotive No. 152 at the head. The station pictured here is the second railroad depot in Fairbanks. The three rail track allowed joint use of both narrow gauge and standard gauge trains on the same rails. The narrow gauge track left the dual gauge track a short distance north of town only to rejoin the standard gauge line in front of the Fairbanks station in Garden Island.
Photo courtesy of the Candy Waugaman Collection

Paragraph 1 references a 1994 conversation between the Author, Nicholas Deely, and M. Maddox, former Tanana Valley Railroad/Alaska Railroad employee

A view of the tracks clearly showing the three rails.
Photo courtesy of the Candy Waugaman Collection

Paragraph 3 references Fairbanks Daily News-Miner, March 21, 1913

Paragraph 4 references Fairbanks Daily News, August 4, 1908

Paragraph 1 references Fairbanks Daily News-Miner, August 25, 1917

Paragraph 2 references Fairbanks Daily News-Miner, August 13, 1917

In the meantime, the system continued to deteriorate. Years of heavy use and minimal maintenance on the ties, made from untreated native wood, took their toll. Accidents were frequent. One of the most notable accidents was that of the "Merino Wreck". As the train approached the trestle near Merino, the coach went off the track, taking the baggage car (a converted boxcar) with it. Both the baggage car and coach No. 204 were totally destroyed. The accident happened, as best as can be ascertained, on May 15, 1916, just before the Tanana Valley Railroad was taken over by theAlaskan Engineering Commission. These two photographs were taken by George A. Parks, a passenger on the train. In 1925, he became the Governor of Alaska.
Photo courtesy of the Jim Brown Collection.

Paragraph 3 references Along Alaska Trails, Pg. 28

was particularly the case with the ever frequent mining accidents — for the operation of a "*Drift Mine*" can best be looked at as an accident waiting to happen. Since any serious injury required an immediate transfer to Fairbanks for treatment in hospital, one of these "*speeders*" would be summoned to hasten to the nearest railhead to pick up the injured individual. The motor car would then speed off to Fairbanks where the victim would be transferred to St. Joseph's Hospital which was just across the street from the depot.

These emergency trips were not without danger nor, as a matter of fact, was the routine operation of these units. While hurriedly passing through a 6 degree curve at Happy Station, on August 24, 1916, one of these motor cars carrying five passengers, struck a telephone pole that had dropped across the tracks. The unit was derailed sustaining considerable damage. Fortunately there were no injuries to the passengers. The simplicity of these units were such that they usually could be repaired the same day, to return in a short time, to their regular schedule.

However, just 11 days prior to this accident a fatal episode *did* occur. John Rolla, a foreman of a section crew on the railroad, lost his life when his small gasoline motor car jumped the track at the "Y" where the Chena line cut off from the Fairbanks - Chatanika main line. He was coming along at a good speed, at about 7:30 P.M., after completing an inspection tour of the Chena line, unaware that a switch was thrown against him at the junction. The motor car jumped the track and did not stop until it had covered a distance of about 50 feet. The impact was so great that the forward motion of the railcar, upon striking the closed switch stand, threw him in front of the run-away unit, which ultimately came to a stop after it had *fallen on top of him*. One of the workers from the Experimental Farm, which was located a short distance from the accident, found the body and immediately summoned help. Wade H. Joslin, Manager of the railroad, came out to the scene of the accident posthaste with the *Electric Car*! Serving as an ambulance, the unit spared no time as it silently hummed, at top speed, to town. Arriving at Garden Island, Rolla was taken, straight-away, across the street to St. Joseph's Hospital. Sadly, he passed away the following morning from complications resulting from a broken neck, a broken back and a broken rib that had pierced his lung and caused it to collapse.

In spite of all its efforts to reduce costs, the railroad was unable to compete with the highway system. This was not only due to the stagnation within the mining industry caused, in part, by government intervention, but also to the high cost incurred in the operation of the line. By 1912, the railroad was operating trains at the rate of only once or twice a week. This led many to suggest that the railroad warning signs at the highway crossings should be removed and replaced with — "*Watch out for the trains on Tuesdays!*"

There was however, for a relatively brief period of time, a ray of hope for the railroad during these twilight years. This came in the form of newly developed antimony, tungsten and quartz mines which were located some distance off the mainline of the railroad. Olnes served as a distribution and receiving depot for several of these mines in the *Tolovana* district — some 50 miles away! Mineral bearing rocks and ore were shipped from the mines by wagon and then transferred to railcars for shipment to one of the several *stamp mills* closer to Fairbanks and Chena. The ore was crushed and the desired elements were removed. While the whole operation at these mills required only the presence of two workers, the need for large amounts of water to flush out the pulverized rock made their continued operation precarious!

During the month of August, 1915, the railroad transported nearly 300 tons of antimony ore to the river for shipment to San Francisco. Monthly gains thereafter had reached such proportions that it was anticipated that the railroad would be transporting, ultimately, as much as 1,000 tons of ore monthly. To retain this newly found source of revenue, the line planned to reduce freight rates again and to build additional spur tracks out to the mines — some as long as five to ten miles in length. None of these plans materialized, for in addition to its inability to handle the present freight demands, it had to continuously contend with the fickleness of nature. These difficulties became evident when on March 24, 1916, blizzards and snowdrifts at Summit severely disrupted train schedules. Tungsten ore from the mines at Gilmore Creek, that was destined for the stamp mill in Chena, a plant partially owned by Martin Harrais, was markedly delayed. The mill had crushed its last batch of stored ore on this day, and tie-ups of the railroad made it impossible to continue work even though there were large amounts of mineral ore still available at the mines. The mines and the mills closed down!

It was the BEGINNING OF THE END for the Tanana Valley Railroad!

Photo courtesy of the Dorothy Wilde Collection

TUNGSTEN STAMPEDE

Word reached town this afternoon that a rich tungsten strike had been made at the head of Dome creek and traced for the greater part of the way over to Cleary creek. George White, the mining engineer in charge of the Eagle antimony mine, is reported to have investigated the matter and to have declared it promising looking.

This afternoon Mr. White was called up on the telephone by a representative of the News-Miner and and asked for information regarding the matter. He replied that he could say nothing at this time. For this reason and from his attitude regarding the matter, it would appear that there was some foundation for the report.

The strike is supposed to have been made by "Missou" and a man named Rick and for the last week a quiet stampede has been in progress.

Fairbanks Daily News-Miner, May 23, 1916.

Fairbanks Daily News-Miner,
March 24, 1916

ANTIMONY MINE IS LARGEST IN UNITED STATES

Fairbanks Will Have the Distinction of Having the Largest Antimony Mine In The Country.

OUTPUT WILL BE 3,000 TONS SURE

First Boat Took Out About 1,100 Tons; Shipments to Go Forward Every Few Weeks for Summer.

Fairbanks Daily News-Miner, May 24, 1916. "Antimony Mine is the largest mine in United States"

CHAPTER 11

THE BEGINNING OF THE END

Uneven track, plus the lack of ballast, allowed accidents to occur frequently. In many cases, trains would just ride over the rails and come to rest on their side. Fortunately, injuries were minimal.
Photo courtesy of the Colin MacDonald Collection

Author's Note:

Alaskan Engineering Commission, establish by an act of Congress on March 12, 1914 (38 Stat. L 305), authorizing the President of the United States to locate, construct and operate railroads in the territory of Alaska.

This Commission became an agency under the executive directors of the Secretary of the Interior. It was entrusted with the survey location, construction and operation of a government railroad in Alaska.

It took over, finished construction and operated the Alaska Northern Railroad (Seward to Kern Creek on Turnagain Arm — 70.8 miles) on March 12, 1914, and ultimately operated the Tanana Valley Railroad.

Further, it operated coal mines, telegraph and telephone lines, docks, a transfer service, power plants, etc.

On May 8, 1914 the Commission gave the order to start the survey for the route which was to become the Alaska Railroad.

From:

The Alaskan Engineering Commission, by Joshua Bernhardt.
D. Appleton and Company • New York • 1922 Copyright by: The Institute for Government Research

"Skookum" — An Alaska / Yukon expression for someone, or something, that is good or great.

DURING THE COMING YEAR WE EXPECT THE GOVERNMENT RAILWAY TO START WORK HERE AND CONNECT WITH OUR LINE, FORMING A CONTINUOUS LINE FROM THE COAST TO CHATANIKA, WHICH FROM THERE WILL BE EXTENDED TO THE YUKON RIVER.

Part of Tanana Valley Railroad ad printed in the Fairbanks Daily Times, December 25, 1915. The entire ad printed on page 89 of this book.

Fairbanks Daily News-Miner, September 16, 1916

From many dissident voices across the territory, there developed a unified clamor for the construction of a railroad from the ice-free port of Seward to the interior of Alaska. Private enterprise initially attempted to accomplish this task — but failed! The Alaska Central Railroad was built up through the rugged Kenai peninsula, while the Alaska Northern Railroad threaded its way along the shores of Turnagain Arm, an extension of Cook Inlet. Both ultimately met with failure. Federal government intervention was demanded by the residents of the territory, to either complete these projects or to develop an alternate one.

Hearing this clarion call from Alaska, President Taft, in 1912, created the *Alaskan Railroad Commission* to formulate a plan for the development of a line from the sea to the Interior. By 1914, the railroad bill was passed by the United State Congress, and the *Alaskan Engineering Commission* was formed to execute the order of the federally mandated railroad. This Commission was formed by the newly elected President Wilson, and was placed under the jurisdiction of the Secretary of the Interior. It consisted of W.E. Edes, Chairman; Thomas Riggs, Jr.; and Lieutenant Fredrick Mears. Thomas Riggs, Jr., in particular, was a real "skookum" individual, well suited for the rigors of Alaska. It could be said that he was a real "sourdough" of the north who had served his apprenticeship as a *voyageur* by raft, canoe and pole boat on many streams in Alaska. Early in his tenure with this Commission, he would travel down to Nenana from Fairbanks by canoe, to over-see the construction of the railroad. Since he was in charge of all the work in the Interior, he elected to make Nenana his home and the headquarters of the new railroad.

Nenana, from time immemorial, had been the gathering place of the Nenana Denayee, the native people of that region. Earlier in this century it was called *Tortella*, white man's corruptive pronunciation of the proper Indian name for this fish camp, *Togottele*. This was also the name given to the hill just across from this fish camp on the other side of the Tanana River. It means, "Something that is floating" — that is, the perceived image that one gets when looking at the hill from across the river. At its base is located a sacred Indian cemetery, long desecrated by the white man!

The residents of this village were, and still are, *Athabascan Indians,* who are closely related to the Navajos and Apaches of Arizona. While they were a migratory people, they did not accept the theory that they migrated to Alaska by coming across the Bering Sea many thousands of years ago. Rather, they felt that they were placed here as an act of creation!

On June 15, 1915, the first survey party arrived in Nenana ostensibly to start the initial phases in the construction of the new railroad. No exact route had been planned as yet, and no money had been appropriated by the Congress. In any case, they set about on July 17, 1915, to build a waterfront dock, 1,000 feet in length, along the south bank of the Tanana River. The prevailing philosophy was to build the railroad, simultaneously, from two directions — both north and south of Nenana. The line to the south was to be constructed to the width of 4 feet 8½ inches, which was consistent with the established North American practices. The line to the north, on the other hand, would start from the opposite side of the Tanana River in North Nenana and proceed through the vast wilderness of the Interior to Fairbanks. This line, however, would be built to the three foot gauge of the Tanana Valley Railroad so that an immediate tie-in to the operations of this line could be accomplished. Since the projected massive railroad bridge across the Tanana River was to be a monumental undertaking, both in time and money, priority was given to this line so that commerce could start immediately upon completion of the track-laying program. Later, upon completion of the bridge, this track would be converted to the standard gauge of the *Alaska Railroad*.

Within three months, this lonely fish camp had sprouted into a community of 1,200 people. It outstripped all other Alaska communities in growth, save for Anchorage. Virtually overnight, warehouses, machine shops, homes, schools and a hospital were built along *planned* streets, which were illuminated by electric street lights.

While all this constructive activity was going on in Nenana, Chena, in 1916, on the other hand, was being dismantled. Strangely, now the buildings of Chena were following the population that had long ago left town. Piece by piece each building was being dismantled for shipment to Nenana where they were, once again, re-assembled. Oddly, it was Martin Harrais, the father of Chena, who had initially furnished the material and money to build most of the town, and was now engaged as the superintendent in the destruction of the buildings. Every time he tore off a board from the frames of old Chena, he could not help but feel that he was committing an indignity upon an old friend.

During the demise of this community, entrepreneurs took advantage of the dying town to reap a handful of cash from its ashes. Perhaps one of the most creative was Jack Smith, from Fairbanks, who went to Chena with the hope of getting a few dollars from the wrecking business. Alert to every opportunity, he saw upon his arrival that there were a great number of men at work, but no one to feed them. With two dozen eggs and two pounds of bacon, he started on his restaurant adventure, charging each customer "One Buck" for a plate of "Ham An." So great was his business that he became one of the most prosperous business men of dying Chena.

In this feeding frenzy of making a "buck" at all costs, even Doc. Laymon, the local physician, got into the act. Apparently there was a shortage of sleeping quarters, for each day, as one building after another was torn down, space available was at a premium in Chena. He appropriated every nook and cranny as sleeping quarters for his clientele, even if it meant sleeping them cross-wise, one along side the other, on a pool table! Hudson Stuck, an Episcopal minister best described Chena's demise as the

NEW TOWNSITE PROBABLE AT NENANA

Nenana Townsite May Be Changed From North to South Side of the Tanana River.

DIFFICULTY IN DRIVING PILES

Frost Was Struck, Making It Necessary Every Night to Thaw the Ground Before Driving Piles.

According to F. H. Bailey, the division engineer in charge of work at Nenana, it may become necessary for the railroad commission to transfer the townsite from where it was surveyed last fall, on the north side of the river, to the south side, near Duke's trading post. The position of the bridge may be changed also from the place originally intended. This will depend, however, upon what the commission discovers when they start boring to learn the formation.

Part of an article printed in the Fairbanks Daily News-Miner, June 2, 1916

Paragraph 2 references Fairbanks Daily News-Miner, September 12, 1916

Paragraph 3 references Fairbanks Daily News-Miner, September 23, 1916 classic Alaskan instance of short-sighted greed that over-reached itself — and lost out! All that remains of Chena today is its nebulous history, a camp ground on the original townsite and the still flowing Tanana River.

Steamboats On The Chena, Pg. 78

In the meantime, old Nenana was now becoming a thing of the past. The Indian people who lived between the Nenana and the Tanana Rivers in harmony with the land were forced to relocate. A stable society was replaced by a new wave of "progress", lawlessness, bootlegging, gambling and most importantly, diseases as yet not known to these people. So desperate became the status of the Indians because of these socially foreign disruptions that they immediately set about to draw up laws for the conduct of *their own people*. Specifically, they mandated rules and regulations for themselves designed to counter the disruptive actions of the white people which they were now witnessing. Foremost among these edicts were the following: no one shall gamble, drink whiskey, sell or give away intoxicating drinks, steal or fight with a friend. Equally important, no woman or child should go into town by themselves! Social problems presented a challenge, for white man's diseases represented almost total annihilation. Out of a community of 190 people, 75 persons perished from influenza during the epidemic of 1918, a proportion far greater than that experienced by the white settlers!

Author's Note:
The Influenza referred to in paragraph 1 was "Spanish Influenza."

If the expectations were great for the immediate construction of the railroad, the actual progress up to this time was not. Although a construction camp and headquarters for the line were being established in Nenana, the actual route for the system was yet to be determined. This issue was in contention as far back as 1914, when it was originally thought that the line should come up along the *south* shore of the Tanana River to Fairbanks. While nothing ever materialized out of this plan, neither did anything else develop, as an alternate plan, until actual clearing of the land occurred in 1917.

Fairbanks Daily News-Miner, May 14, 1917

Paragraph 3 references Fairbanks Daily News-Miner, November 24, 1916

To compound matters further, the Alaskan Engineering Commission, which was responsible for the construction of the line, had not been allocated any money by the federal government to carry out their mandate. The Civil Sundry Bill, which represented the financial allotment in question, still awaited passage by the Congress well into May of 1917! Local residents throughout the Interior were so up in verbal arms over this procrastination that Thomas Riggs, one of the members of the Commission, attempted to placate them by stating, "The construction of the railroad between Nenana and Happy would be an easy matter under any circumstances. If necessary, this could be completed in 30 days." Ironically, all this was said even before the route of the new railroad had been plotted.

Never-the-less, contracts were being let out to private developers for brush cutting along the right-of-way *which as yet had not been truly defined*, and for the production of 150,000 railroad ties plus 1,000,000 feet of logs. These logs were to be used for the construction of bridges, trestles and quarters for the railroad workers.

HERE'S GOOD NEWS

WASHINGTON, D.C., May 19.—(Delayed in Transit)—The sundry civil bill, which includes $6,247,000 for the construction of the Alaska railroad, was reported in the House today. The total amount of appropriations provided in the bill is $127,000,000. It is expected that this bill will be passed through congress within the next week. This money will become available on July 1.

Fairbanks Daily News-Miner item printed May 20, 1916

In addition, orders were placed for rails, which by September 13, 1916, were arriving in Nenana. This load consisted of several hundred tons of rail which came into town on board the river steamer *Tanana*. Such was the frenzy at this time of the year to beat the freeze-up, that many of the steamers actually found themselves racing each other up the Yukon and Tanana Rivers to Nenana to unload their cargoes as quickly as possible, so as to avoid being weathered in for the winter. As might be expected, this could and did, quite often, lead to catastrophic results. Such was the case on September 20, 1916, when the riverboat *Atlas* elected to race the riverboat *Tanana*. The *Atlas* ran into a sandbar in the river and sank at 5 P.M. that day. There was a heavy loss of material destined for the prospectors and trappers — but no loss of life!

Boat after boat continued to come up the Yukon and Tanana Rivers with loads of rail and other construction material during the autumn of 1916, with no assurance that the government would allocate money for this project. Some of it was unloaded in Nenana, while the rest was trans-loaded to smaller river craft, out of deference to the low water of the Tanana River, and sent upstream to Fairbanks and Chena.

During the same period in September that all these supplies were arriving in Nenana, the Alaskan Engineering Commission was busy in Fairbanks leasing land for eventual railroad construction. Twenty acres of ground were obtained, on the north side of the Chena River, from the Tanana Valley Railroad. The site was almost in line with the present day St. Matthews Episcopal Church. A river dock and warehouse were to be constructed there.

In the meantime, September had passed into October and still there was no real work carried out on the construction of the railroad. Sensing the extreme silent frustration that was building in the community because of the lack of progress on the line, the Alaskan Engineering Commission quickly moved to avert a public relations disaster. If for no other reason than to be symbolic, the Commissions elected to lay some track. On October 21, 1916, the whole town of Nenana turned out to witness the driving of the *first spike* into the *first railroad tie* of the *first section of permanent track*, in the Interior, on what was to become the *Alaska Railroad*. The honor was bestowed upon a Mrs. James Duke, whose primary credential for this role was that she was the first non-Indian woman to reside in Nenana. It took her *18 blows* to drive in the spike. Other than bringing in a steam shovel to muck about in the mud while filling in cuts along the roadbed, this represented the sum total of work carried out on the railroad in 1916.

In spite of all the urgency to get the railroad built, only 4,356 *feet* of narrow gauge track had been laid between the end of 1916 and April of 1917. Shortly after break-up on the Tanana River, on April 29, 1917 at 1:50 P.M., track construction was resumed again. An additional 19,630 feet of track was laid

Paragraph 1 consolidates references Fairbanks Daily News-Miner, September 13, 1916, and Fairbanks Daily News-Miner, September 20, 1916

Paragraph 3 references Fairbanks Daily News-Miner, September 11, 1916

Paragraph 4 consolidates references Fairbanks Daily News-Miner, October 21, 1916 and Fairbanks Daily News-Miner, November 18,1916

PRESIDENT WILL ANNOUNCE THE ROAD ROUTE TO FAIRBANKS NOW

Executive Order Will Be Issued Fixing the Location of the Government Railway Between Nenana and Fairbanks—Will Be Based Upon the Report of Tom Riggs Which Tells the President of New Discoveries in the Tolovana and Kantishna Districts—Money Will Not Be Paid the Canadian Bondholders for Some Time. Because of Suits Brought.

WASHINGTON, D. C., June 30.—Railway Commissioner Riggs will submit to President Wilson and Secretary Lane a report of railway conditions in Interior Alaska upon which the president will issue an executive order fixing the location of the Alaska government railway route from the Nenana river to Fairbanks. Commissioner Riggs in a preliminary report already received makes reference to the new gold discoveries in the Tolovana and Kantishna districts, which are added reasons why the railway should be constructed and at once.

Headline and a portion of an article printed in the Fairbanks Daily News-Miner, February 18, 1915. Unfortunately, "Now" was not to be anytime soon.

Fairbanks Daily News-Miner, August 26, 1917

JOSLIN ROASTS CONSERVATION

VALUABLE PROPERTY IS RUINED BECAUSE OF CONSER-VATIONISTS.

Upon arriving at Seattle, on this present trip to the States, after having visited many places along the coast, including Katalla, and territory contiguous thereto, Falcon Joslin, in an interview in the Seattle Times, "roasted" the conservationists for the property that was ruined by the withdrawal of coal lands. The article will be of interested to many Fairbanksans who were interested in coal, and to others who know Mr. Joslin. It reads:

"There is no doubt as to the mineral value of the country, he said, both as to coal and to oil. The coal makes the hottest fire I ever saw. It burns white. Some oil drillers left a 12-inch steel bit in an open fire while they went to lunch. When they returned the piece of solid steel had melted. Coal that will melt steel will melt copper ore.

"They are pumping three wells and making 15,000 gallons of gasoline a month, while about 30,000 gallons of kerosene and other illuminating oils are allowed to go to waste, because they have no means of conserving it. The plant is a small 2-by-4 affair that cost about $2,000. The parafine and lubricating oil around the plant is a foot thick on the ground. The little plant manufactures all the gasoline used in that part of the country. Only one man and a boy are drilling. They are without money to buy proper equipment, but are getting along as best they can.

"I never saw a sadder sight in my life than the former camp of the so called English Coal company. The company put up fine, comfortable buildings, built miles and miles of trails and roads, but the whole thing is deserted and going to wreck and ruin. Ten or twelve miles of railroad track has become two streaks of rust.

"All of this waste was wrought by the withdrawal of the coal lands and the fight put up by the conservationists against the granting of patents.

Fairbanks Daily News-Miner, January 15, 1916

during the ensuing weeks for a grand total of *23,986 feet*. Since it was 47.6 miles from Nenana to Happy, the junction point of the tie-in with the Tanana Valley Railroad, it became evident that this "progress" could not meet with the needs of the people of the Interior. Wood, the primary source of energy for the Fairbanks region, was becoming very scarce and farther and farther away from any supply at all. Even more irksome to these folks was the realization that there was a plentiful supply of coal just south of Nenana sitting, literally, next to the projected main line of the new railroad.

To alleviate this energy shortage, construction of the narrow gauge line to Fairbanks was given the highest priority by having rail construction crews working simultaneously, north from Nenana, and south from Happy. What made this work so imperative was the availability of hundreds of acres of forest along this route which could meet the energy needs of the Interior. Although the track was of a three foot gauge, the roadbed, bridges and curves were built to standard gauge specifications. The rail was laid dead center on eight foot six inch standard gauge ties, but at a width of three feet. Conversion of the track, at a later date, would amount to nothing more than spreading the rail apart to 4 feet 8½ inches, the width of the standard gauge track of the Alaska Railroad.

This eventual tie-in with the Tanana Valley Railroad was imperative for it opened up the whole of the Interior to rail transportation. River movement of passengers and freight upstream from Nenana to Fairbanks and Chena was, at best, an uncertain adventure. Water levels were not static; but rather, they were in a continuous state of fluctuation, hiding dangerous sandbars which lurked just below the surface of the water. As a result, channels were in a constant state of flux. To meet this challenge, freight from larger riverboats had to be transferred to smaller craft in Nenana to raise the draft of the vessel so as to insure passage upstream.

By July, 1917, extensive brush cutting and grading of the roadbed had taken place all along the Goldstream line. This was followed by heavy ground work whereby the draws and muskeg were filled with rocks and gravel. On the northern end of the line, which was now proceeding southward from Happy, priority and urgency fueled the progress. Bob Crawford and Bob Mahar, two of the contractors on the project, were sent out periodically to oversee the construction. On one occasion, they spent a continuous two weeks out in the bush working 12-16 hours a day. With limited foodstuffs available to them, their diet consisted *only* of *beans* and *bacon*, with no coffee, tea, bread or sugar. *It was reported that they had no problems with the mosquitoes*!

It was the intention of the Alaskan Engineering Commission to build at least five miles of track south from Happy before October 9. This plan was predicated on the availability of rail, which, at least initially, turned out not to be the case. On September 17, 1917, however, the riverboat *Seattle No. 3* arrived in Chena towing two barges loaded with rail and other construction supplies. This turned out to

Community frustration with the slow progress of the Alaska Railroad often found its voice in the local newspapers. Bigotry was so flagrant during the early part of the 20th century that derogatory racial, social and ethnic references were commonplace. Not only were these attitudes verbalized, but they were also freely printed by the press.

OUR RAILROAD COMING UP

Alaska Railway Commissioner Arrives in Seattle and Busies Himself Saying Nothing Much.

SILENCE IS A VIRTUE

The Man Who Talks Too Much Will Not Be Able to Get Far Ahead With the Present Government.

Fairbanks Daily News-Miner, April 18, 1915

THERE'S A NIGGER IN THE WOODPILE - SOMEBODY LIED

The Man Who Introduced the Alaska Railroad Bill In Congress and Who Has Fought Conscientiously and Consistently for Its Passage Is Assured by Secretary Lane That the Railroad Will Be Built to Fairbanks—Lane Does Not Plan to Start Work From Nenana North—Edes and Wickersham Will Urge Him to Do So Plans Are Made to Start Work From Nenana South Toward Coal Fields—Knockers Will Please Note.

Signal Corps. United States Army.
Telegram.

Received at:
25 V X 50 NPR
P CS WASHINGTON DC DECEMBER 13 1915
NEWS MINER
FAIRBANKS ALASKA

Secretary Lane's estimate contains two hundred twenty-five thousand dollars for clearing and grading twenty-four miles from Nenana south. On Monday the sixth I strongly urged that work north from Nenana be started. Am awaiting Commissioner Edes's arrival when will again urge it. Lane said to assure people that road would be built to Fairbanks. Erwin saw Lane Friday Knocking. No influence but beg Fairbanks quit knocking.
4:35 pm WICKERSHAM

Fairbanks Daily News-Miner, December 14, 1915

WILL WORK BOTH NORTH AND SOUTH IN SPRING

Delegate Wickersham Is Assured That Road Will Be Built to Fairbanks—Edes Is In Favor of Consructing Both North and South From Nenana in the Spring—Wickersham and Alaska Engineering Commission In Harmony and Working Together—That Spells Success for Fairbanks—Is Encouraging News.

The following telegram was received by the New-Miner at 2:30 this afternoon:

"Long innterview with Chairman Edes today urging equal appropriation of the work both north and south from Tanana Crossing. I am confident that the work will proceed both north and south from crossing of Tanana in the spring.

"Both Lane and Edes strenuously assured me that the plan included the completion of the road to Fairbnks. Looks all for the best from this end.

JAMES WICKERSHAM"

Fairbanks Daily News-Miner, December 23, 1915

"SEE ALASKA FIRST" IS NEW R.R. SLOGAN

Railroads Throughout the United States have Started Campaign for Tourists for Alaska—Tons of Literature on the North Is Being Circulated Now—Boat Traffic Is Expected by American Yukon Navigation Company and Alaska Steamship Compny—Washington, D.C., Is Besieged With Tourists Booklets.

WASHINGTON, D.C., Feb. 18—Never, since the purchase of Alaska has there been such a movement of foot to secure tourist trade for the north that there is this year. Throughout the east and particularly the national capiptal, tons of literature is in circulation, setting forth the charms of the northland as a paradise for tourists. Practically every leading railroad of the country has united in the movement, with the slogan, "See Alaska First" Excellent inducements are being offered by the transcontinental railway comanies and it is believed that several thousand people will make the trip north during the summer.

In addition to the trips across the country, the steamship companies are advertising attractive trips through the Panama canal, with a visit at the San Diego fair, and a railroad journey from there to Seattle, and thence to Alaska by boat.

Some of the companies are simply advertising a trip along the southern coast of Alaska, but the majority of booklets tell the prospective traveler that to see the north as it is, to see the country that is to be opened up by the building of the government railroad, it will be necessary to make the trip down the Yukon and up the Tanana valley thence out by way of St. Michael and Nome, and back around the southern coast of Alaska.

A. Y. N. Helping

SEATTLE, Feb 18—The American Yukon Navigation company is carrying on the most persistent advertising campaign in its history, according to officials of the compapy, as it wants to induce all tourists to make a trip through the Interior. Manager Wheeler said that the company would be able to handle all the tourists that are to make the trip and spepcial efforts will be put forth to give them the best accommodations possible.

The campaign of the eastern railroads is in conjunction with the A. Y. N. efforts to show tourists the wonders of the north. Mr. Wheeler says, and his company has done a great deal of work in prepapring statistics and in issuing booklets describing the places of interest to visit.

Without doubt the tourist traffic in the north during the coming summer will be greater than ever before Mr. Wheeler says, and he advises the hotel men to get ready to receive the guests.

On the other hand, premature enthusiasms may have caused their own troubles.

be the first shipment of rail to the northern end of the line. This rail was immediately dispatched to Happy via the Tanana Valley Railroad for use on the new line. After laying 2½ miles of track out of a projected five miles, this batch of rail was totally utilized. To determine what progress this construction effort had accomplished, the officials of the Alaskan Engineering Commission elected to inspect this new track for themselves. On October 8, the first "train", under the direction of Wade Joslin, headed down the track using the old Beech electric car. They departed from Fairbanks and traveled down to Happy. Here the switch from the main line of the Tanana Valley Railroad was thrown to bring the entourage onto the new line. Of the 2½ miles of track that was constructed, they inspected only 1½ miles of it — apparently to their satisfaction. Earlier in the month, a siding of 200 feet was installed at Happy in anticipation of the great amount of railroad supplies and other construction material that was now going to be diverted to the Happy-North Nenana line.

Fairbanks Daily News-Miner, September 17, 1917

Fairbanks Daily News-Miner, October 9, 1917

Fairbanks Daily News-Miner, October 1, 1917

With revenues declining and operating costs approaching a point of insolvency, Falcon Joslin, in the meantime, directed his attentions toward the oncoming Alaska Railroad, hopeful that it would represent the salvation for the Tanana Valley Railroad. It was his desire that the government railroad would terminate in Chena, for this would allow the Tanana Valley Railroad to survive as the transportation vehicle between Chena and Fairbanks. Earlier, he had been negotiating with the Alaskan Engineering Commission relative to the prospect of it either purchasing the entire Tanana Valley Railroad or that portion of the line which would parallel the government line from Happy to Fairbanks. Initially, the Alaskan Engineering Commission declined this offer. In 1916, the government railroad instead, started grading operations on a line north of what was to be the University of Alaska — a line parallel to the Tanana Valley Railroad. Since the Tanana Valley Railroad already had railroad facilities in Fairbanks, the Commission rethought its original position on this matter and came to the conclusion that there was no point in building the parallel railroad line.

Author's Note:
Tanana Valley Railroad's decline in revenues was directly linked to the decrease in gold mining in the creek areas, which declined due to high costs of labor and materials, scarcity of fuels, and low purchasing power of gold during World War I.

Negotiations continued on through the spring of 1917. On June 2, 1917, Secretary of the Interior Lane announced that the Alaskan Engineering Commission was authorized to purchase the entire property of the Tanana Valley Railroad at a price not to exceed $300,000. In the meantime, until this transaction was formalized, it was authorized to operate this line. However, before this transaction could be carried through, there were some legal hurdles to be overcome.

On Friday, July 27, 1917, the Columbia Trust Company of New York, holder of the first mortgage on the railroad, initiated a suit to foreclose the mortgage on the Tanana Valley Railroad. The forerunner of this company was the Knickerbocker Trust Co. of New York which had put up the original money for the Tanana Valley Railroad to purchase the Tanana Mines Railway. Besides a principal of $782,820, the railroad owed back payments of interest due to the bank since 1914. As there seemed to be little chance that this interest would be paid, the bank felt that the railroad was insolvent, and accordingly asked for

Fairbanks Daily News-Miner, July 29, 1917

Fairbanks Daily News-Miner, November 1, 1917

Author's Note:
There were three reasons why the Alaskan Engineering Commission benefited from purchasing the Tanana Valley Railroad. First, by buying the lines, no new track had to be built. They could simply add a third rail to the already existing line, running 7.3 miles, from Happy to the end of the government rails at the terminal in Fairbanks. Second, adequate terminal facilities were already in place and operating in Fairbanks and no new stations needed to be built. And last, it was thought that enough mining potential existed around the Fairbanks area to make it become profitable again for a railroad, should the price of gold raise enough or the Nenana Coal fields make sufficient fuel cheaply enough available to encourage the mines to reopen.

Paragraph 1 references Fairbanks Daily News-Miner, July 20, 1960

Author's Note:
Spell of the Yukon by Robert Service presents an excellent reflection of the conditions present during these times and the mind set of the men who lived through them.

a foreclosure. This action was taken so that the title of the railroad could be cleared before it passed to the ownership of the government.

On October 1, 1917, the Fourth Division Court rendered a decision in favor of the Columbia Trust Co. seeking a payment of $782,820 in First Mortgage Bonds against the railroad. At two o'clock in the afternoon on November 1, 1917, at a Marshal's sale in front of the courthouse in Fairbanks, the railroad was sold by Deputy Marshal H.R. Tull to A.R. Heilig, an agent for Benjamin L. Allen, for *$200,000.* Benjamin L. Allen of New York was *President* of the Columbia Trust Company and *one of the Vice Presidents of the Tanana Valley Railroad*! The Marshal's Deed was issued on November 14, 1917 — and thus, with the title cleared, the Tanana Valley Railroad was sold by Benjamin L. Allen for *$300,000* to the Alaskan Engineering Commission, on December 31, 1917. The price came out to be approximately $6,818 per mile. From this moment, until the line became the Alaska Railroad in 1923, it was operated by the Alaskan Engineering Commission. It was now public property.

The men who built this railroad were subjected to a regimen of unbelievable hard work and were forced to exist under the most primitive living conditions. Generally, they were mostly European immigrants. There was no smiling camp steward to direct these new arrivals to their quarters, for there were none to be had! One had to *provide for themselves and build their own accommodations*. Two types of quarters were in vogue at that time. Tents (without floors) and pole bunks covered with wild hay for mattresses, represented a quick deterrent against the northern elements. The other alternative, which was generally found along the winter access trails or tote roads, were crude log houses that were hastily constructed. They were chinked with moss so as to make a tight fit between the irregular logs. The roofs on such houses were made of strong poles, with very little pitch. They were then covered with bark, hay and moss. This was further covered with two or three feet of topsoil or earth. Door hinges were cleverly made from bent nails or from old, leather boot tops. Homemade wooden latches were then secured to the doors to keep them shut. Locks were not necessary, for at this time, Alaska was a land without locks or policemen!

These trail accommodations were a combination cookhouse, storeroom, dining hall, social room and bunkhouse. The bunks were built out of poles placed in tiers across both ends of the room — similar in arrangement to post office boxes. Measuring four feet in width, and eight feet in length, they were referred to by the "track gangs" as "muzzle loaders". The extra two feet at the end of the bunks was for personal storage. It was not uncommon to find four tiers of such bunks placed in six different rows at the end of the room. Crude ladders were placed along side their "muzzle loaders" to allow the men access to the upper bunks. Two oil lamps were suspended from the ceiling to provide some illumination, while a centrally placed stove provided some heat and a place of warmth for social interaction. Conditions were so primitive and cold that it was a feeling among the workers that in order to be awarded the coveted title of "Sourdough", one had to spend at least a night in one of these Alaskan "muzzle loader" bunkhouses!

Author's Note:
It is interesting to note that the majority of the hand work done on the construction of the Alaska Railroad was carried out by Scandinavians, Russians and the Irish. Generally they were, as yet, non-naturalized immigrants who endured the worst of conditions. Laborer numbers peaked in 1917 at 4,500 workers.

Except for the 2½ miles of track constructed south from Happy, 1917 seemed to be the year which was primarily devoted to the task of procuring and transporting railroad equipment to Alaska. Seemingly endless boatloads of cargo churned up the Tanana and Chena Rivers to inundate the docking facilities at Nenana, Chena and Fairbanks. In addition, thousands of tons of freight were sitting on the docks in Whitehorse for shipment into the Interior. These materials came north by ship from Seattle to Skagway. From there the freight was transported to Whitehorse by the White Pass and Yukon Railway. It appeared that the chief purchasing agent for the Alaskan Engineering Commission, C.E. Dole, went on a shopping spree in the states to such a degree that it took several ship loads to deliver the materials to the Interior! This manifest consisted of 700 tons of light rail, switches and rail fittings, in addition to several flatcars. *None of this material was new* for it had just recently been in the service of the U.S. Engineering Corps. on a jetty construction project near Fort Canby in California. The rail was to be used to repair some of the bad track and roadbed on the Tanana Valley Railroad, and act as temporary track for the new construction on both the narrow and standard gauge line that was coming up from Broad Pass.

Fairbanks Daily News-Miner,
August 4, 1917

With the onset of the autumn of 1917, the rivers began to freeze up and the Interior towns gradually withdrew into their winter solace. For all practical purposes, all construction on the railroad stopped. The region went into a deep sleep only to be awakened in the new year to face a nightmare the likes of which the world has seldom seen!

The year of 1918 struck with vengeance — an epidemic of influenza laid a mantle of death, pain and suffering upon the whole world, including Alaska. Deaths, across the board, were staggering. In Alaska, all the people were affected, but especially the Native Americans. Among these people, the onslaught was not limited just to influenza, but also included measles, smallpox and scarlet fever. Close to one third of the Native American population of Nenana became victims of these diseases.

Laying track on the new Nenana extension from the banks of the Tanana River toward Fairbanks. This photo was taken in the vicinity of the Ohio Roadhouse, in 1920. Consolidation No. 151 is in the background.
Photo credit: Anchorage Museum of History and Art. Acc. #BL 79.2. 4883, of the Alaska Railroad Collection

Quarantine stations were set up between many of the communities of the Interior. No longer could unrestricted travel take place. Such stations existed on the outskirts of Fairbanks and Nenana. One wishing to travel to either of these towns was "invited" to spend ten days of sequestration at either one of these holding zones before being allowed to proceed into either one of these respective communi-

Influenza pandemic of 1918
Excerpts taken from the Fairbanks Daily News-Miner, November 8, 1918.
The Influenza Menace . . .

Nome, November 7, 1918 ---"There are about 300 cases of Influenza here. Three white people are dead and the Eskimos are *dying like rats*. Seventy Eskimos have died to date, and the Influenza is spreading along the coast. The sick Eskimos had no one to keep up the fires for them or to give them medicine and they were found frozen in their beds. A number of Eskimos have been rounded up and put in a big building where they can receive attention, but several of them, after being placed in the building, *hanged themselves*. Most of the business are closed and the hospitals are all crowded."

NOME NUGGET

Fairbanks . . . "Health Officer Sunderland is establishing quarantine stations to prevent the disease from reaching this camp. The Piledriver roadhouse is being opened today as a detention hospital, and all who try to come to Fairbanks will have to serve time there . . . Nenana will shut herself up Monday next and none will be able to leave there."

"Last night a special meeting of the City Council was held . . . . It was decided to take all precautions necessary, the City guaranteeing the bills if the Territory Government won't pay them . . . . and to fumigate the mails . . . People who move to the Interior (should) be stopped and turned back . . . (Nobody) need have any urgent desire to come to Fairbanks at this particular time, unless (to) come away from the disease . . . We don't need more inhabitants now, under the circumstances. We can live off ourselves until health conditions improve Outside . . . . We don't even need the mail . . . Those who receive *fumigated mail* will be less anxious to receive mail after they have opened one paper fumigated with some vile-smelling thing, (as) they will (get) a *violent headache* . . . Until a case arrives in Fairbanks (there) will be plenty (of) time to think of closing the schools and public places. Take away anything which would attract (people's) attention away from disease and trouble, and herd then in (to) their homes with no place to go and nothing to THINK about but INFLUENZA, and they (might) develop influenza through their thoughts always being there-on . . . But, it wouldn't be bad business to compel all people . . . to KEEP AWAY FROM FAIRBANKS until the influenza is conquered. Fairbanks doesn't need influenza. *We have all the troubles we need without it."*

ties. As a result, a short visitation to any town would require a total of 20 days of isolation — ten days at the community to be visited and ten days upon return to one's home. Somehow, during this great period of misery, pain and death, work on the extension of the line from Happy Station to Nenana, continued.

Shipping season had started again in earnest, and by July 4, 1918, the river steamer *Shusanna* had made a second trip from Nenana to Fairbanks, bringing with it another load of rail and other track construction material plus the *first load of coal* for the Interior. This coal originated in the Nenana coal fields. Included further in the freight manifest was another narrow gauge locomotive No. 151, a 2-8-0, Consolidation, some additional flatcars and some dump cars to be used for track construction. Although locomotive No. 151 was to be the largest power unit on the narrow gauge, it did not see much service. Apparently it was too heavy for the poor track conditions of the Tanana Valley Railroad. All of this cargo was unloaded in Chena, which by this time was already beginning to see the end for itself.

Fairbanks, Daily News-Miner,
July 20, 1918

Appropriations for the construction of the railroad still had not materialized, even at this late date in 1918. As a result, progress continued to be limited, much to the dissatisfaction of the public. To raise the moral of the people, the Alaskan Engineering Commission elected to put on a Potlach to help elevate their spirits. With much derision, the *Fairbanks Daily News-Miner* reported the upcoming event by stating, "The Alaskan Engineering Commission wanted to celebrate their success *in being a failure*. They want to put on this Potlach for the people of Nenana to celebrate what *little* they may have accomplished in building the line a relatively short distance, north from Nenana, into the woods."

Autumn of 1918 passed, uneventfully, into the winter and spring of 1919. There was now a feeling of great expectation that the line to Fairbanks would, at last, be completed. Although the influenza epidemic had not as yet abated, financial appropriations were being made to construct the narrow gauge line to the north. During the preceding months, the Alaskan Engineering Commission instituted a sequence of priorities for the construction of this new line. The new, large, steel bridge across the Tanana River *would not be built until the standard gauge line came into Nenana from Seward*. The narrow gauge line going north from North Nenana, and coming south from Happy, was to be the primary focal point of construction. In the meantime, a contingent of small riverboats would ferry both passengers and freight back and forth across the Tanana River and up-stream to Chena and Fairbanks.

The need to build this line as soon as possible was due to the wood famine which had developed in Fairbanks and the Interior. Factors responsible for this acute shortage were two-fold, availability and cost. During the first year of logging in 1903, extensive forests surrounded the whole of Fairbanks. So great had become the need for energy, as the years passed by, that the trees had been cleared for a mile from each side of the Chena River bank and for as much as a 100 miles upstream. The Northern Commercial Company itself needed 15,000 cords of wood a year to manufacture enough electricity to meet the demands of Fairbanks and the mines. As the availability of local wood decreased, the costs of wood increased. By 1915, the cost for a cord of wood had risen to $11.25. Had coal been available, it would have cost $5.00 a ton and would have released the same amount of energy as *two cords* of wood. Not only had the U.S. Government prevented the mining of coal on federal land, but it had elected, also, to impose a tax of 25 cents for each cord of wood logged on government land — which meant all of Alaska. Falcon Joslin ridiculed this tax as an utter folly of deception. He alleged that the cost involved in collecting the tax would be higher than the revenue returned.

Along the course of the Nenana extension through the lower Goldstream Valley were extensive strands of virgin timber. Five miles of new main line, which would act as a temporary spur track, were laid as quickly as the availability of rail allowed. To accomplish this, Frederick Browne, Construction Engineer for the Northern Division, compromised basic railroad engineering principles by ordering that the new track be built directly on top of the *quivering muck* which underlaid the moss and brush along the route. This was done with the realization that gravel and ballast were not immediately available to stabilize the track. At a later date, he reckoned, the track would gradually be rebuilt to established standards once the desperately needed fuel was rolling into Fairbanks.

Fairbanks, Daily News-Miner, September 24, 1918

Paragraph 2 consolidates 3 different references: Fairbanks Development Quarterly, Vol., 3, Number 2, Summer, 1992; *Steamboats on the Chena*, pg. 81,& *Ghost Railway in Alaska*, Duane Koenig, Pacific Northwest Quarterly, January 1954, pg. 11

Huge amounts of wood were used on a daily basis by local businesses and homes in Fairbanks and the creek areas. Virgin stands disappeared at a rapid rate. Sleighs hauling as much as 7 cords of wood at a time were regularly seen along the winter trails.
Photo credit: University of Alaska Fairbanks, Alaska and Polar Regions Department. Acc. #89-166-383, in the archives, of the Albert J. Johnson Photograph Collection

NEEDING TIMBER THEY TAKE IT

In Order That the Alaska Government Railway May Have Plenty of Timber for Construction Work and Other Uses, the President Today Declares Timber Reserves in Great Variety and Extent in Alaska, Mostly in the Coast District. Where Timber is Timber.

Fairbanks Daily News-Miner, June 16, 1915

Railroad in the Clouds, pg. 79

Information in top paragraph was condensed from a letter written by John Raap - Special Disbursing Agent, Alaskan Engineering Commission, on April 19, 1918 to Mr. W. C. Edes - Chairman, Alaskan Engineering Commission and consolidated with information gleaned from Fairbanks Daily News-Miner article printed October 10, 1919.

The timber was cut into four foot lengths and loaded onto the narrow gauge flatcars which had high racks built up at each end to contain the wood. Each car carried ten cords of wood and were coupled together to form a whole train consisting of newly-cut logs. They were immediately hauled over the new railroad to the junction at Happy where they went by the Tanana Valley Railroad to Fairbanks and the mines. A total of 275 cars were shipped out to these respective localities during the winter of 1918-1919. More wood could have been hauled out from these logging camps had there been enough flatcars available to accommodate this tonnage!

Even after the new railroad was completed to Fairbanks and the shipment of coal to this destination had started, the use of wood continued for many years. It was quite common to see long strings of *standard gauge* flatcars, laden with logs, still heading for Fairbanks. After transloading these four foot logs to the narrow gauge cars of the Tanana Valley Railroad, this cargo continued on its way to Chatanika.

From 1914 until the beginning of 1919, much, and very little, was done to finally complete this new narrow gauge line. It wasn't until mid-year of 1919 that the Congress finally got about to make the financial allotments necessary to pay for the railroad merchandise that it had already accumulated throughout these years. Yet, in spite of the continuous armada of riverboats that came into Nenana, Chena and Fairbanks with all kinds of supplies, never was there any rail to complete the job! Finally, during the month of August, 1919, the riverboat *Tanana* came upriver with a barge in tow loaded with enough rails to complete the track construction from the end of steel at Happy south to North Nenana. These rails were unloaded at the dock in Chena and then transferred by the Tanana Valley Railroad to the end of track south of Happy. By then, the rest of the projected right-of-way had been cleared of brush, graded and made ready for the newly arrived rails. Wooden ties were positioned, rails were centered on them in their proper places — followed by the rhythmic, piercing shrill sounds of steel pounding on steel as the "gandy dancers" (track layers) drove home their spikes to form the new track.

RAILS ARRIVE FOR OUR ROAD

YET A LITTLE WHILE AND THE CARS WILL BE RUNNING HERE FROM COAL MINE

The steamer Tanana on her last trip from Hot Springs brought with her a bargeload of rails in tow, which were delivered at Chena, where the cargo was transferred to the Tanana Valley railroad for ultimate delivery at the end of steel south of Happy Junction. The Tanana's tow, which is one of the two barges of steel brought up the Yukon by the steamer Seattle No. 3, represents the last of the rails needed to complete the government railroad between Nenana and Fairbanks.

The Tanana also brought some oil, which was unloaded at the oilhouse on the Island opposite Nenana, and some other freight for the Northern Commercial company.

Fairbanks Daily News-Miner, August 23,1919

So great was the emotion in the people of Nenana over the prospect that they could leave the town at last, that immediate plans were made for a mass exodus to Fairbanks. In contrast to the miners on the creeks, these people lived an isolated life for Nenana was nothing more than a large construction camp! They couldn't come to Fairbanks for parties, dances or any of the other cultural events which were so prevalent in that city at the time. While there was some river travel between the two towns, true freedom of movement was not there. One could now visit all those great shops in Fairbanks and socialize with friends who only a short time ago were so far away and distant.

Finally, on Tuesday, November 4, 1919, the first "Thru train on the North Nenana - Garden Island Transcontinental Railroad," as it was facetiously referred to by local newspaper reporters, came into town with a large crowd of people from Nenana. The fare was $3.50 round-trip. A similarly enthusiastic bunch headed south from Fairbanks to North Nenana. From there it was just a short trip across the Tanana River by boat to Nenana. However, this was not a "thru" train in the truest sense. What occurred was that one train left North Nenana for Fairbanks, while another train departed from Fairbanks to North Nenana, both meeting at *Standard*, approximately half way along the line. At this point the passengers were obliged to leave one train and transfer to the other train, for as yet, there was still a sizeable gap between the north and south sections of track. Trudging by foot through the bush, passengers boarded the other train which reversed its direction and headed back to its point of origin!

On November 7, 1919, the two segments of track on this narrow gauge line were finally joined together. The train could then travel directly from North Nenana to Fairbanks. This new section of track was operated, for the time being, as part of the Tanana Valley Railroad. The whole entity north of Broad Pass was to become known as the *Northern Division* of the *Alaska Railroad.*

Except for the bridge which was to span the Tanana River, the only large project that needed to be completed was the standard gauge line which was coming north from Broad Pass to Nenana. What was holding up this progress was the lack of rail that continued to plague the railroad. In its search for this commodity, the attention of the Alaskan Engineering Commission became focused on the status of the Chena line of the Tanana Valley Railroad. By 1919, the line to Chena was to see only one train a week — with it now originating and returning to Fairbanks. It did operate special trains to Chena however, as the situation called for it. Periodically, for one reason or the other, riverboats would come up the Tanana River to Chena and proceed no further. In conjunction with these riverboat arrivals, special train runs to Chena were made for the benefit of the in-coming passengers and freight. Ultimately, however, it was during this year of 1919 that the last merchandise was to be landed in Chena!

To meet this acute need for rail, the Alaskan Engineering Commission, on or about March 29, 1920, removed a spur track that switched off from the Chena line in Chena. It extended from the Northern Commercial Company store to the "Y" at the local saw mill. Since there was still considerable freight to be hauled from Chena to Fairbanks, no further track removal was contemplated at that time.

Not a month later however, the "gandy dancers" were back, removing the Chena line *in its entirety*. All available flatcars were sent to Chena to bring up all the freight and wood that still remained. The crew from bridge and building No. 5 removed the rail on approximately April 23, 1920, which was immediately sent south for use on the line coming up from Broad Pass.

FIRST TRAIN COMES THROUGH

This is a great day in Our Town.

Today the first "thru" train on the North Nenana-Garden Island transcontinental railway made town in good time, bringing passengers from Nenana and taking passengers to Nenana.

This train didn't go all the way through. The two trains met at Standard, half way, changed partners and the engine from here brought the train from North Nenana here. Friday, however, the train will go all the way and back again, leaving here in the morning, making North Nenana and returning here in the evening.

Those coming on the train state that the completion of the road to here has caused the builders to lay off 250 workers, and that many of them are hiking for the Outside, but that is another story—now that we have $6,000,000 more for railroad work than we had yesterday.

Fairbanks Daily News-Miner, November 4, 1919

Paragraph 1 references Fairbanks Daily News-Miner, November 4, 1919

Paragraph 2 references Fairbanks Daily News-Miner, November 7, 1919

Paragraph 4 references Fairbanks Daily News-Miner, March 20, 1920

Paragraph 5 references Fairbanks Daily News-Miner, April 16, 1920

CHAPTER 12

STANDARDIZATION OF THE NENANA LINE

Locomotive No. 152 at the station in Fairbanks. The station house is the second railroad depot in Fairbanks. Note the three rail track which allowed use of both standard gauge and narrow gauge trains on the same track.
Photo courtesy of the Candy Waugaman Collection

T.V. RAILROAD IS REHABILITATED

NENANA, Aug. 10.—Word has just been received in the office of the Alaskan Engineering commission of the departure from Anchorage of S.K. Ward and party, via the Broad Pass. They are bound for the Interior end of the Government railroad for the purpose of rehabilitating the Tanana Valley railway.

An inspection of the roadbed of the railroad recently proved conclusively that it would not stand up under the demands of heavier rolling stock and it will be the duty of Mr. Ward, who will act in the capacity of Locating Engineer, to rebuild all bridges, roadbed and structures and to put them in proper condition. The road will not be rebuilt for standard gauge equipment at present, the Commission having shipped in a considerable amount of rolling stock of narrow gauge to handle the traffic demands between Nenana and Fairbanks until such time as the bridge across the Tanana shall have been completed. This cannot be done for some time as it is practically impossible to secure bridge steel with the present condition of the steel industry, when practically the entire output of the country is demanded for war industries.

With the completion of the railroad between Nenana and Fairbanks it is proposed to handle all business by means of a ferry across the anana at this point, the ferry being of sufficietn size to handle several cars at one time.

Fairbanks Daily News-Miner, August 10, 1918

Paragraph 2 references Fairbanks Daily News-Miner, April 16, 1920

Paragraph 3 is condensed from a series of letters written from May 15, 1924 to April 22, 1924, by J.T. Cunningham - Superintendent of Transportation, Alaskan Engineering Commission to H. Horn - Superintendent of Tracks, Bridges and Buildings, Alaskan Engineering Commission.

No sooner had the Nenana extension been completed than the scenario for its conversion to the standard gauge was set into motion. Sadly, it was the beginning of the end for one of America's northern- most narrow gauge lines. During the period of four years that this extension existed, from 1919 to 1923, some of the most colorful operations of any railroad took place. The unique operations were created by the need to operate a railroad of two different gauges, separated by the Tanana River and a bridge that as yet had not been built!

To integrate the operations of the narrow gauge lines of the Northern Division, the Tanana Valley Railroad utilized the Nenana extension as if it were an integral part of its own system. It would run the trains out to Chatanika and then down to Nenana on alternate days from Fairbanks. The whole operation was run, theoretically, with two locomotives. These locomotives consisted of No. 152, a new 4-6-0 Ten Wheeler which had arrived in Nenana piecemeal and was assembled there in the shops of the Alaskan Engineering Commission. Ultimately, this locomotive became the workhorse of the line. The other unit was No. 151, a 2-8-0 Consolidation. Unlike all of the other locomotives of the Tanana Valley Railroad, which were wood burners, this unit was converted into a coal burner. This conversion had not only an economic advantage, but also saved two or more hours from the total time that was needed to load wood into the tender for the run to Nenana. It was first placed into service in December of 1919 and was used almost exclusively, on the North Nenana run.

Oddly, locomotive No. 151, which was one of the most powerful engines on the line, was prohibited from use in the summer on the steepest grade on the railroad — that is, between Gilmore and Olnes. It was the feeling of J.T. Cunningham, Superintendent of Transportation, that the tracks of the Tanana Valley Railroad were in such a deplorable state that they could not stand the weight of either locomotive 151, or equally so, that of No. 152. Engine No. 52, the 2-6-0 Mogul, which was lighter than the other two locomotives, was pressed into service over this, the most difficult portion of the "mountain line"! It epitomized the "Little Engine That Could," for it was just barely able to haul the train over the hill. Come winter, however, the roadbed was frozen in position which then offered the desired stability to the track. Locomotives No. 151 and No. 152 were then put back into service. The former ran to North Nenana over a well constructed and firm standard gauge roadbed, but on narrow gauge tracks, while the latter held down the Chatanika run.

Though there were many unique arrangements to "bridge" the gap across the Tanana River, nothing could even begin to approach the *piece de resistance* of northern railroading in the winter. Once the river became frozen, the narrow gauge Northern Division *would lay temporary track across the ice.* A spur track was constructed off the main line in North Nenana and descended down from the north bank of the river. It then crossed over the ice to the foot of the boat dock in Nenana. Narrow gauge trains from the north and standard gauge trains from the south came within a few feet of each other, but on

During the summer months the tracks of the Tanana Valley Railroad were considered to be too unstable to support the heavy 2-8-0, locomotive No. 151 which belonged to the Alaskan Engineering Commission. However, it was soon realized that the frozen Tanana River, which usually had at least 48 inches of ice, could not only support this locomotive with its train, but also an additional locomotive — at the same time!

New narrow gauge passenger car being delivered to the A.E.C. government railroad in Nenana. Tracks and flatcars in the foreground are resting on the frozen Tanana River.

Passenger trains crossing the frozen Tanana River from Nenana to North Nenana. Tracks are resting directly on the ice. Note the 0-6-0 shunter locomotive going in the opposite direction.

Transferring freight from the narrow gauge A.E.C. freight cars to waiting sleighs. Train rests on the *ice* of the frozen Tanana River in Nenana.

With the aid of a chute, freight and baggage is transferred in Nenana from the standard gauge boxcar, on the upper level, to a narrow gauge boxcar sitting on the frozen Tanana River below.
Photo credit: University of Alaska Fairbanks, Alaska and Polar Regions Department. Acc. #84-75-286N, in the archives, of the Col. Frederick Meais Collection.

different levels — the narrow gauge on the frozen river while the standard gauge rested on terra firma! Just before break-up, the track would be pulled up from the ice to be reused the following winter. The first time that this technique was used occurred on the morning of February 19, 1919 under the direction of track foreman, Thomas K. Blakely. Not only did the *gandy dancers* spike the rail into waiting ties across the river, but they also wrote a new chapter into the history of railroad engineering. To this very day, Russian railways periodically make use of this procedure on their northern branch lines.

The impetus for an ice crossing of the railroad was due to the dire need for an additional source of energy in the Fairbanks region. Many voices were heard advocating the need to transport coal from the Nenana mines directly across the ice. Hearing this call, the railroad stated that it was willing to do this contingent upon a commitment by the consumers to purchase a specific amount of coal. To this, the residents of Fairbanks gave a mute answer. By implication, therefore, this meant that no commitment would be given. This need for a minimum load was predicated on the costs incurred to lay, and reclaim, the track each winter. Although these orders from the public to transport coal into the Interior never materialized, the railroad also had a great need for this coal. Sensing the prevailing stalemate, a couple of entrepreneurs, Fred Parker and August Conradt, accepted the challenge. They were in possession of a couple of motor trucks which could haul the coal, resting in the *standard gauge* hopper cars which had been recently delivered to the rail yards in Nenana, across the ice to the waiting *narrow gauge* cars on the north bank of the river. Twenty-four, four wheeled dump cars were filled and dispatched to the north on the next train. This coal was to be delivered to the stations along the line for use by the railroad.

Fairbanks Daily News-Miner,
February 17, 1921

Ultimately, the demand for coal became so acute that the railroad, unilaterally, elected to lay light track across the ice so that it might be able to use its own gondola cars to transport larger quantities of coal. The railroad continued this operation until the ice across the river could no longer bear the weight of the train. It still continued to drop off some of the coal at wayside stations, but the larger amount of it was unloaded at a stockpile in the Tanana Valley Railroad yards in Garden Island. By then the energy crisis had become so critical that the coal assumed the role of a precious commodity to be auctioned off, when available — at the drug stores!

Paragraph 3 references letters exchanged January 31, 1922 between J.T. Cunningham, Trainmaster, Alaskan Engineering Commission and F.A. Hansen, Engineer, Maintenance and Construction, Alaskan Engineering Commission

During the winter of 1920-1921, locomotive No. 151, the large Consolidation, was used in service, for the first time, *across the ice* on the Tanana River. Since it was a very cold winter that year, the ice became very thick enabling it to support the weight of this heavy engine. However, during the winter of 1922, this was not to be the case. Instead, a smaller and lighter 0-6-0 engine, No. 830, was pressed into service for this ice crossing. In time, locomotive No. 151 was to be used sparingly, relegated only to yard shunting and work train service.

With the subsequent demise of this locomotive for use in mainline service, only locomotive No. 152, the Ten Wheeler, was available for use on the whole narrow gauge system. Any malfunction of this unit resulted in a severe disruption of the service provided by the railroad. This was to happen on a particularly cold day during the month of January of 1921 when the scheduled freight train stopped to pick up some additional southbound cars at Happy. This locomotive, which was pulling ten loaded freight cars, already had its horsepower compromised by the frigid weather and *leaky boiler tubes*. Thus the added nine cars which were coupled to the train at Happy ultimately stressed the engine to such a degree, that it promptly "expired" upon its arrival in Nenana! Since the locomotive had to be repaired before it could go north again, the cars and the train crew that were needed for the run up to Chatanika were not, consequently, available for duty. As a result, the whole line was forced to shut down!

During the winter of 1921-1922, passenger rail service was carried out across the ice between North Nenana and Nenana for the first time. Concern, however, about the integrity of the ice to support one of the regular locomotives of the line led to the use of a "dinky" 0-6-0 type engine to shuttle the passenger car back and forth across the frozen Tanana River. This locomotive with its attached passenger car(s) became known as the "North Nenana Limited"! After crossing the ice, passengers would board a waiting train, on either side of the frozen river, to continue on with their journey.

In time, after confidence was gained by the railroad in the feasibility and safety of this operation, the regular narrow gauge passenger train would run right down the north bank of the river and onto the ice. After making this crossing, the train would pull in at the bottom of the river dock of Nenana — still resting on the ice! Standard gauge cars which had come north from Seward were positioned *on the dock* directly above the narrow gauge train. Freight which was going north to the Interior, was transferred down to the narrow gauge cars below on the ice via a chute. It usually took two narrow gauge cars to hold the contents of one standard gauge car. Other cars were loaded directly from horse-drawn wagons and sleighs which had pulled up on the ice along side of the narrow gauge cars.

Passenger transfer was much more of a cumbersome ordeal. After traveling for two days from Seward, passengers bound for the Interior were required to disembark from their train in Nenana at 6:45 A.M. They were then obliged to make do as best as they could until 1:30 P.M. when their narrow gauge train would depart for Fairbanks. Although the train was scheduled to arrive in the Interior at 5:50 P.M., this was more an illusion than a reality!

By 1922, the railroad maintained a twice-weekly service along the entire length of its line from Seward to Fairbanks. Trains departed Seward on Mondays and Thursdays, arriving and departing Nenana on Tuesdays and Fridays for the scheduled four hour and twenty minute trip to Fairbanks. Southbound trains left Fairbanks at 8:00 A.M., on Tuesdays and Fridays, to arrive in Nenana at 12:20 P.M. It wasn't

Fairbanks Daily News-Miner,
January 12, 1921

TRAIN DID NOT RETURN

The Nenana train, which left on schedule yesterday morning, was unable to return to Fairbanks last night owing to leaking boiler tubes it became necessary to delay the return trip until the necessary repairs could be made. According to the latest information it will leave Nenana at 1 p.m., today, and reach town shortly before 6 o'clock this evening.

The leaky tubes, it is believed, were due to overloading the engine during a drop in temperature when it becomes a strain on the locomotive in picking up the cars that have been standing heavily loaded for any length of time. The train left Fairbanks with 10 cars, and picked up 9 additional at Happy, making 19 in all, which is considerable load on the Happy line.

Fairbanks Daily News-Miner, January 12, 1921

ON TONIGHT'S TRAIN

FOWL
DUCKS
GEESE
TURKEYS
TANSY
SPARERIBS
FINNAN HADDIE
VEAL
EASTERN OYSTERS
OLYMPIA OYSTERS

Fairbanks Tri-Weekly News-Miner, November 22, 1921

THRU TICKETS DISCONTINUED

In a wire to Agent Palmer, J.T. Cunningham, Trainmaster, at Anchorage, issues the following instructions today: "On account of the fact that the ice over the Tanana river is dangerous, discontinue selling tickets for stations where it is necessary to cross the river, until further advise. Passengers are to take their own chances on getting over rivers. Will advise as soon as ice moves, so that service can be resumed."

Both of these articles reference
Fairbanks Daily News-Miner, May 1, 1922

NO PERISHABLES FOR OUR TOWN

Them perishables as was coming to Our Town on last night's train ain't. Which is something else again.

They are having trouble in getting freight across the ice at Nenana and perishables are their principal trouble. So the Railroad will sell OUR perishables in NENANA at the best price they will pay the loss to the shippers.

They are figuring and scheming to run a cable across the river and cross freight by the aerial train route and they may get to that but not in time for this shipment of perishables.

LATER–The local merchants have signed release for damages from freighting the perishables across the river at Nenana and the Railroad will start work of trying to get them across and to Our Town

Paragraph 3 references Fairbanks Daily News-Miner, May 1, 1922

Paragraph 4 references Fairbanks Daily News-Miner, May 12, 1922

until 8:00 P.M. of the same day that the southbound train departed for Seward. As usual, passengers, in the interim, were left to shuffle for themselves.

When the ice was solid and intact, railroad operations functioned rather smoothly. However, with spring break-up, operational decisions were determined on a hour to hour basis, pending the status of the ice. When break-up came in 1922 on April 8, passengers going south from Fairbanks were notified that their tickets would be honored only as far as North Nenana. They were requested to leave their coach and directed to walk *across the ice* to Nenana, for it had become too dangerous to allow the train to make the trip across the frozen river. Upon arrival in that town, they were to rebook passage for their on-going trip to Seward.

While the ice was still relatively intact freight could be, to some degree, transported across the river in wagons. Once the ice started to break-up, it became a transportation nightmare. Regardless of the ice conditions on the Tanana River, heavily laden freight trains, which were co-ordinated with ship arrivals in Seward, were dispatched to North Nenana. Since the river could not be crossed, mountains of merchandise piled up on the docks. This tonnage included heavy construction material, hardware and staple food; but too, a large amount of gourmet delicacies such as fowl, ducks, geese, turkeys, tansy, spareribs, Finnan Haddie, veal and Eastern Olympia oysters — foods which were also served to the passengers on the train.

With such a larder virtually rotting away under the noses of potential customers, the railroad periodically would sell these perishables in Nenana for whatever price the market would bring. The differential loss would be reimbursed to the shipper by the railroad. To possibly eliminate this bottle neck, there was some consideration given, at one time, to string a cable directly across the river so as to expedite the transfer of freight during the break-up. Nothing ever became of this concept. In the meantime, freight cars continued to arrive in Nenana, compounding an already bad situation. Upwards of 18 cars at a time could be found in the yards, collecting demurrage and choking off the terminal.

Although the river had to be free of ice to allow this merchandise to be transferred to the other side, some of the most dangerous and challenging crossings were being made in the interim. *Riverboat* men, for a fee, would be most happy to jeopardize your life, their life and the "life" of their boat, in the attempt to make the crossing for you. For a fee of $5.00 per person in the daylight and $10.00 per person at night, these river men would thread their craft through the icy, ever changing leads, while attempting to avoid massive floating icebergs. Any contact with them would, most assuredly, send both passenger and boat to the bottom of the river. Even after a successful passage, no one arrived on the opposite shore dry! Once the river was free of ice, which was usually by the middle of May, boat service was used to bridge the gap. The riverboats *Matanuska*, *Midnight Sun* and the *Sun Flower* were commissioned to transport passengers and the ever-mounting freight resting on the docks in Nenana.

As a result, with the onset of the summer of 1922, the primary goal of the railroad was to complete the biggest challenge that it had on the entire system — a railroad bridge across the Tanana River at Nenana! By August, 1922, Commissioner Mears of the Alaska Engineering Commission felt that the progress made, up to that point, on the construction of the bridge was so great that it exceeded his most sanguine hopes for an early completion. In the meantime, while all this construction was going on, summer had passed into autumn and then winter! Ice began to form making the river unsafe for navigation. By November 13, the ice, in certain spots was considered sufficiently strong enough to permit a safe crossing. Teamsters were madly scouring about looking for the "bridge" that would allow them to get across. When the ice was finally considered safe, mail, express, baggage and freight were transferred across the frozen river by horse-drawn teams. Similarly, teams were provided for women, children and invalids for their crossing; males were on their own to get across the best way that they could!

On November 15, 1922, the track across the *false-work piling* of the new steel bridge under construction was completed. Trains from the south now connected with the narrow gauge lines on the north side of the river. With this completion, the excitement, the danger and the uncertainty of cross-river, ice and water passage came to an end! The standard gauge track was pushed up to the Moss home in North Nenana, which was just south of the "Y" at the station. The station was abandoned that morning, and Agent Keck's stint at this post also came to an end. The railroad entered its final stages of a new beginning!

With the final completion of the 701 foot long Mears all steel bridge, one of the longest single spans in the world, trains began to come into North Nenana from Seward at a steady pace. It was completed during the early part of February, 1923, and stood high above the Tanana River so as to allow unrestricted navigation on the waterways. Standardization of the narrow gauge lines now became top priority. Two separate and different methods were employed to accomplish this task, while at the same time retaining the ability to use both the standard and narrow gauge systems simultaneously.

The most immediate change took place in the North Nenana railroad yards where it now became necessary to accommodate the trains of two different gauges. This dual service was accomplished by laying a third rail parallel to and along side of, the narrow gauge track, thus making is suitable for use by both systems. This same procedure was used in the transformation of the narrow gauge line between Happy and Fairbanks.

Regauging the main line north to Fairbanks was relatively easy. Initially, the rails were laid dead center on standard gauge ties. Thus, it was a simple task for the *gandy dancers*, as they worked north, to

Paragraph 1 references Fairbanks Daily News-Miner, August 7, 1922

Paragraph 2 references Fairbanks Daily News-Miner, October 10, 1922

Author's Note:

Although the completion of the Alaska Railroad spelled the beginning of the end for the steamboats, it is interesting to note that the Alaska Railroad owned and operated five steamboats on the lower Tanana and Yukon Rivers. The first two boats were the *Gen. Jeff C. Davis*, and the *Gen. J.W. Jacobs* purchased from the U.S. Army in 1923 to replace the Alaska Yukon Navigation Company, which ceased operations on the Tanana and lower Yukon Rivers in 1922. Both boats were retired in 1933 and replaced by the *Nenana*. This boat was built at Nenana in 1922, and purchased by the Alaska Railroad in 1923. She served 22 years and was retired in 1954. *Nenana* was donated to the Fairbanks Chamber of Commerce and moved to Alaskaland in 1967. She has been refurbished and is open to the public. Also owned by the railroad was the *Minneapolis (renamed Benny K.)* purchased in 1927 and retired in 1947, the *Alice II* purchased in 1927 and retired after W.W. II, and the *Yukon*, purchased in 1942 and retired 1946.

Author's Note:
President Wilson authorized the Alaskan Engineering Commission to purchase supplies and equipment in the conventional manner, to utilize materials available along railroad right-of-way, to transfer surplus equipment from the Panama Canal and (later) to transfer war-surplus equipment under the acquisition Act of 1914. The Commission paid for the repairs and transportation of the items, but, on the other hand got them free-of-charge, resulting in a savings of approximately 50-60% over the cost of new materials.

remove the spikes from the ties, spread the rails of the 3 foot gauge track to 4 feet, 8½ inches, and then to respike the rails into their new position. The simplicity of the whole procedure proceeded so smoothly that by April 26, 1923, the work teams had arrived in *Saulich's* wood camp, which would later become milepost 450.5 on the Alaska Railroad. For a short while this location became the transfer point on the railroad for the two systems of track. Upon arrival at this "break-of-gauge", passengers were obliged to muck through the soggy tundra — dragging, pushing or carrying their luggage as they transferred from one train to another. Freight, on the other hand, was reloaded on the other train by the railroad crews. Once everybody and everything had been transferred, the trains would reverse their direction and return to their point of origin — thus enabling the passengers and freight to complete the journey.

By May 8, 1923, the break of gauge had come as far north as Happy, milepost 463.0 on the Alaska Railroad. Since this was the junction with the Tanana Valley Railroad, the Alaska Railroad also assumed their mileage identification which, in this case, was milepost 7. This junction was to become, temporarily, a very busy transfer center for passengers and freight, for it was here that the Tanana Valley Railroad diverged from the Nenana extension to proceed off to the east to Chatanika.

Fairbanks Daily News-Miner,
May 8, 1923

On the evening of May 7, 1923, at 10:30 P.M., the first standard gauge locomotive arrived in Happy. It was a 2-6-0 Mogul, No. 618. This engine originally worked on the Panama Railroad during the construction of the canal. At the time that the locomotive was purchased by the Alaskan Engineering Commission, it had a gauge of 5 feet but was regauged to the 4 feet, 8½ inches of the North American standard. With its arrival, Happy became an important junction, transfer point, and in time, the beginning of double gauge operations into Fairbanks. A station with a telegrapher was opened on the same day that locomotive No. 618 arrived in Happy. It was Agent Hazel's job to co-ordinate and direct all freight train movements. This transfer of passengers and freight from Chatanika to *Seward* bi-passed Fairbanks altogether. At the head of a mixed consist, locomotive No. 618 departed Happy, south bound, after meeting with the narrow gauge creek train. It was the first train to travel to the coast without any interruptions from water barriers (Tanana River) or break-of-gauge in the track! After crossing this river via the newly constructed *Mears* bridge, the train went on to Curry where it made an overnight stop before proceeding on to Seward.

Ironically, by the time locomotive No. 618 had reached Happy, there was no rail available to complete the change of gauge operation toward Fairbanks. By May 16, only one mile of track east of Happy had been converted to standard gauge. So desperate was the situation that some of the sidings were deliberately taken up so as to make rail available elsewhere along the route of construction. By May 29, enough rail had been scrounged up so as to bring the three rail track to the west side of Dead Man's slough.

Spanning the slough was a small post trestle that fell victim each spring to the destructive forces of ice flows. Pilings and bents were splintered and torn away from their moorings to leave either a greatly weakened bridge or no bridge at all. Simultaneously, while it was being destroyed by the forces of nature, the bridge and building crew was hard at work rebuilding it. To minimize damage, the railroad came up with an ingenious method to counter-act the destruction. They would cut out, from the base of each pier, about a foot of ice. This procedure would loosen the anchor ice, thus decreasing the stress on the bridge. As a result, damage to the bridge was lessened. In any case, the trestle had to be totally rebuilt so as to accommodate the heavier standard gauge locomotives and cars.

By the first week of June, 1923, Garden Island had been reached. Overnight, the general population in and about the depot increased by the hundreds. Almost immediately a new tent city arose in the terminal yards which was dubbed as, "Steeseville-On-The-Track"! No sooner had the standard gauge track arrived on the scene than stakes appeared in the ground all over the terminal. These were surveyor's markings set in the soil to guide the construction that was to take place. Plans were made for a new depot and for the immediate destruction of the old Tanana Valley Railroad roundhouse and storage sheds. Standard gauge tracks were being laid all over the place at the expense of the narrow gauge track. Ultimately, this reconstruction led to one of the saddest days; on Thursday, June 14, 1923, even the hard, grizzled faces of the old timers wept! Not only was the narrow gauge track pulled up from along the side of the depot, but plans were made to relocate the Tanana Valley Railroad depot across the tracks to the west. It was converted into a dormitory for the railroad people. This resulted in the destruction of last vestige of the Tanana Valley Railroad in Fairbanks.

On the other hand, there was some redemption for all these changes that were taking place. It came in the form of the "universal equalizer" — money! Each new railroad worker brought in from one to ten paychecks to an already sagging economy. Almost equally important, but in another sense, the construction brought in additional people to play *baseball*; for, from the standpoint of sports, this was what this town was all about.

With the completion of this new track into Garden Island, which eventually become a part of Fairbanks proper, the work of the Alaskan Engineering Commission came to an end. The whole railroad network including the Tanana Valley Railroad, became the **ALASKA RAILROAD**. On July 15, 1923, a sweltering 90 degree day, President Harding drove in the "Golden Spike" at the north end of the Mears bridge in Nenana. Alaska's railroad was officially completed!

The rails will probably be taken from the crew stationed, at bridge No. 5, and will be sent down the line early during the next week. The first of the steel to be removed will be shipped to a point a short distance south of the Nenana canyon, where light steel is an immediate necessity.

Excerpt taken from article printed in the Fairbanks Daily News-Miner, April 16, 1922

"STEESEVILLE-ON-THE-TRACK" ON THE MAP THIS MORNING

THE TENT-TOWN WHICH IS CALLED THAT SPRANG UP IN A NIGHT, AND BY TONIGHT WILL BE TENANTED BY AN HUNDRED RESIDENTS.

"Steeseville-on-the-Track" arrived last night, and today across the River just a nice walk from town is a city of tents which tonight will house *(together with the cars which come in today for housing purposes)* an hundred "steel" workers who are here to complete the Alaska Railroad to Our Town. As the days go by, this number will be augmented. With the bridge repairs completed 60 carpenters will arrive, for building purposes, and they will be added to by workers of all kinds who may swell the number to 500. And, **ALL WINTER LONG** there will be a payroll over there, of railroad workers.

Fairbanks Daily News-Miner, June 14, 1923

Paragraph 3 references Fairbanks Daily News-Miner, June 22, 1923

Author's Note:

Three U.S. presidents were involved in the construction of the Government railroad. President Taft recommended the construction of the railroad in 1913. President Wilson choose the route in 1915. President Harding commemorated the completion of the railroad in 1923. A few weeks after President Harding drove the Golden Spike, the government railroad was officially christened the "Alaska Railroad" by the Secretary of the Interior, Hubert Work.

Local Train Schedule

THE ALASKA RAILROAD

Trains leave and arrive Fairbanks as follows:

7:00 a.m., Tuesdays and Fridays for Seward and intermediate points. Returning arrives Fairbanks 9:45 p.m., of same days. Trains stop overnight at Curry.

8:00 a.m., Monday, and Thursday for Chatanika, returning arrives Fairbanks 3:30 p.m. same date.

8:30 a.m., 12:15 and 4:00 p.m., Monday, Tuesday, Wednesday, Thursday and Friday, Motor Car for College and return. 8:30 a.m. and 12:15 p.m. Saturday. No service Sunday. On return trip from College Motor Car leaves there at 8:45 a.m., 12:30 p.m. and 1:15 p.m.

Fairbanks Daily News-Miner, May 19, 1923

RAILROADING AS USUAL IS ASSURED FOR NEXT WEDNESDAY

THE BREAKS IN THE RAILROAD WILL BE REPAIRED DAY AFTER TOMORROW AND THE ROAD WILL BE RUNNING AS USUAL NEXT WEDNESDAY.

ANCHORAGE, June 14,—News-Miner, Fairbanks—The repairs to the damage on the main line north of Curry have been progressing much more rapidly than was expected. The most serious breaks, including the Nenana River bridge, will be repaired by Saturday, June 16th. The line will be opened and traffic resumed thruout its length on Wednesday, June 20th.

STEESE

Fairbanks Daily News-Miner, June 14, 1923

EMPRESS

Announcement

ON ACCOUNT OF THE WASHOUT ON THE RAILROAD IT IS IMPOSSIBLE FOR US TO RECEIVE FILM AND WILL BE COMPELLED TO CLOSE THIS THEATRE UNTIL FURTHER NOTICE.

Fairbanks Daily News-Miner, June 15, 1923

Narrow guage Alaskan Engineering Commission (Tanana Valley Railroad) train loading passengers and freight at Nenana. It is resting on the frozen Tanana River and will cross over it on its way to Fairbanks. Locomotive No. 151 was considered too heavy for duty, during the summer, on the unstable tracks of the Tanana Valley Railroad, but not too heavy to cross the ice in Nenana during the winter.
Photo courtesy of the Anchorage Museum of History and Art, Alaska Railroad Collection, BL 79.2.1309.

CHAPTER 13

1930 —The End

The Fox "School Train". Residents of the creek towns relied on the railroad to provide many varied services. This photo was taken sometime in the late 1920's.
Photo courtesy of the Olga Steger Collection

Author's Note:
The Alaska Railroad cost nearly $60 million in federal appropriations and required 8 years of work. World War I slowed construction and war-time inflation raised costs, but in the end, the railroad cost about $125,000 per mainline mile. While the railroad operated from Seward to Fairbanks from 1923 on, it was in no sense "complete". From 1923 to 1938 it operated in the red. The railroad celebrated in 1927 when the deficit fell to under one million dollars for the first time. In 1938 the railroad operated for the first time, *in the black* (not counting depreciation charges which were omitted from the accounting). It was also the first year that the railroad had not required Congressional appropriations, and none have been needed since.

Serving a sum total population of 5,400 people along the line from Seward, Anchorage and Fairbanks, a distance of approximately 500 miles, with no great mining bonanza to feed it or provide the tonnage needed to meet its transportation potential, financial problems soon plagued the Alaska Railroad. The U.S. Congress was particularly vehement in its admonition of the General Manager of the railroad to cut costs of its operations. Among the people who lived along the mining creeks, it was felt that the congressional stand was primarily based on the desire to abandon the narrow gauge Chatanika branch line, using the general financial plight as a trojan horse to scuttle "their transportation lifeline"!

Unfolding at this time was another development which even today has compromised marginal railroad operations — the development of a parallel highway system. Further, the nature and character of the freight that was generated on the branch line was not constant but sporadic. Inbound freight was primarily fuel for the mining companies, while outbound freight was miniscule — just a string of empty box and flatcars. Over the parallel feeder road, gas trucks could deliver freight and passengers directly and timely, right to the door of each mining claim, whether it was located right on the railroad line or some distance from it. Already by 1914, 20 trucks or automobiles were running over the old wagon trails and the newly developing roads.

Looking at this predicament, but from a different standpoint, the ultimate tilt toward highway and truck transportation for anything but tonnage, becomes upon analysis, self evident. Freight arriving from Seward over the *standard gauge* Alaska Railroad into Fairbanks had to be transferred to the *narrow gauge* cars of the Tanana Valley Railroad. Once out in the creek settlements, this same freight had to be reloaded onto trucks for their final destinations. For the relatively small amount of freight to be carried, the truck could, with a single transfer at the Fairbanks depot, more conveniently carry what available freight there was directly to the final destination — albeit at a somewhat higher cost!

Alaska Weekly, May 30, 1930 and
Fairbanks Daily News-Miner,
February 2, 1931

From April 1, 1929 to April 1, 1930, the total revenues from the operation of the Chatanika branch was only somewhat over $17,000, as against expenditures of $62,000, making an operating loss of approximately $45,000 — or at a ratio of *four* to *one*. Looking at it from another way, the Chatanika branch lost money at the rate of $121 a day! By the same token, the rest of the Alaska Railroad had operating revenues of $1,165,910.50, with an operating expense of $2,090,212.50, an operations loss at a ratio of TWO to ONE! Thus, this operating loss of four to one, aggravated by the existing freight being shifted to the trucks, made the Chatanika branch the prime target for abandonment.

The initial steps for the closure of this line were made by General Manager Smith of the Alaska Railroad in 1928. His original plan was to close the line for the winter, for it was the period when their operating expenses increased dramatically. While this decision was being formulated, Col. O.H. Ohlson had become the new general manager of the line. All the political pressures that had been mounting for the abandonment of the line, working in concert with the reality of the financial losses and negligible patronage, were placed on his shoulders. Ultimately there proved to be at least four ultimatums from him relative to the closure of the line. The last was as final as death.

Initially, on October 1, 1929, it was announced that there would be no train service to Chatanika during the months of January to April 1, 1930. Winter storms and massive snow accumulations over Ridgetop, made it very expensive to keep the bi-weekly mixed train operating. However, closing the branch line during the winter made no sense, for it was exactly the time that the miners most needed and depended upon the railroad service. Had he considered the issue from the standpoint of public service, he would have kept the line open in the winter and not the summer. Service however, was not the issue; but rather, it was economics.

Even after one of the prime sources of freight for the line, the Fairbanks Exploration Company, had completed its movement of heavy material over the railroad, which took place during this period of evaluations, no appreciable boon in revenues was realized by the railroad. In actual fact, the Fairbanks Exploration Company, which operated five dredges in the Chatanika Valley, very much wanted to see the railroad abandoned for it claimed that the track of the Chatanika Branch of the Alaska Railroad was an obstacle to the company's dredging operations. In June of 1929, it went so far as to have the track relocated between Fox and Gilmore. This involved about 2,815 feet of track and was due to the fact that the Fairbanks Exploration Company was about to chew up that portion of land with its enormous dredges. Ultimately this was the fate of old Chatanika itself, for this company with its dredges, caused such mayhem to the ground that barely a trace of this bit of history is left behind.

This closure, however, was aborted when a petition, signed by 122 people from the region, was presented to General Manager Ohlson of the Alaska Railroad requesting that the line remain open during this expressed closure date. He agreed to their request and rescinded his order with the admonition that the railroad should receive their business not only in the winter but also in the summer. His own words to this affect were, "When I received the petition, my answer was that if we were to continue to operate this line, we expected to get the business all the year, and not turn it over to our competition in the summer, as in the past." Unfortunately, with the arrival of spring, the freight again was diverted from the railroad to the trucks — an act which would seal the fate of the Chatanika branch.

During this period of uncertainty, in May of 1930, as to the viability and ultimate survival of the railroad, necessary maintenance of the line was not only being carried out but was still a high priority. High water between McNeer's and Fox had undermined the track and caused a non-alignment of the line. Crews were dispatched out to this Goldstream locale and the problems were immediately addressed and repaired.

Paragraph 5 on page 138 (which concludes at the top of page 139) references *Railroad In The Clouds,* Pg. 197

Paragraph 1 consolidates references Fairbanks Daily News-Miner, October 1, 1929; and *Railroad In The Clouds*, Pg. 197

Paragraph 2 consolidates references *Railroad In The Clouds*, Pg. 197; and *The Alaska Railroad*, Pg. 577

Paragraph 3 references Alaska Weekly, May 30, 1930

Paragraph 4 references Fairbanks Daily News-Miner, May 20, 1930.

Line Operated Under Heavy Deficit; Will Stop Service June 1.

GENERAL MANAGER DECIDES ON STEP AFTER CONFERRING WITH MINING COMPANY OFFICIALS AND WITH COMMERCIAL ORGANIZATION—WILL RECOMMEND HIGHWAY BE KEPT OPEN.

Because of the large deficits incurred in its operation General Manager O.F. Ohlson of the Alaska Railroad will recommend to Secretary of the Interior Wilbur that service on the Chatanika brand line be suspended.

He made this announcement this morning before departing for the coast and after thoroughly discussing the situation with all interested parties.

Colonel Ohlson will recommend that service over the narrow gauge between Fairbanks and the creeks be suspended June 1 and at the same time recommend that the Alaska Road Commission receive additional funds to enable it to keep the Steese highway open to Chatanika the year round.

Fairbanks Daily News-Miner, May 13, 1930

Paragraph 2 references Fairbanks Daily News-Miner, May 13, 1930 and June 25, 1930

Paragraph 3 references *Railroad In The Clouds*, pg. 197

Paragraph 4 quote references Fairbanks Daily News-Miner, June 25, 1930

Paragraph 5 references Fairbanks Daily News-Miner, June 25, 1930

Paragraph 6 references *Railroad In The Clouds*, pg. 198

Ohlson Rescinds Order Shutting Down Road To Creek Points July 1

Branch Road Will Be Kept Open Until Such Time As Truck Roads Are Built From Steese Highway—General Manager's Decision Is Announced At Meeting Fairbanks Commercial Club

In response to a resolution by the Fairbanks Commercial Club and a petition signed by residents of the creeks, General Manager O.F. Ohlson announced last night that he had rescinded the order calling for the closing of the Chatanika branch railroad line on the first of next month.

The announcement was made at a meeting of the executive committee of the Commercial Club General Manager Ohlson said that in view of recent information given him concerning the condition of wagon and truck roads leading from the Steese highway he would recommend to Secretary of the Interior Wilbur that the Alaska Railroad continue the operation of the branch line until such time as the roads are improved sufficiently to take the place of the railway.

Fairbanks Daily News-Miner, June 25, 1930

With the virtual "abandonment" of the line by the local community in favor of the highway system, General Manager Ohlson recommended to the Secretary of the Interior that services on the Chatanika branch be suspended on June 1, 1930. At the same time, he recommended to the Alaska Road Commission that additional funds be allocated to the Steese Highway so as to keep it open year round.

The June 1 decision was rescinded, and again the Chatanika branch received a reprieve. However, this was only another postponement of the inevitable — the new date of closure was advanced to July 1, 1930. This date too, was withdrawn on June 24, several days before the time of its execution.

Closure of the branch was delayed for two reasons. The first was pressure from the local community in the form of a signed petition against the closure of the line. However, closer scrutiny by Ohlson revealed that out of the 114 signatories, only two could be considered to have hardship cases which might be incurred by the closure of the line. He considered the rest less than legitimate for they had obvious interests other than the survival of the line.

The second was that the Commercial Club of Fairbanks resisted once again, the closure, insisting that there was a great need for the line in any future development of the region. It promised the railroad that a determined effort would be made to route all business, whenever possible, via the railroad. It exhorted the community to do likewise. To encourage full participation by the community the club published the following resolution: "That the Alaska Railroad is entitled to the full support and co-operation of the entire Fairbanks community, and particularly, for the active support of all the people residing, operating, or conducting any kind of business along the right-of-way of the Chatanika branch line or said road, or contiguous thereto, and that all such people should patronize the railroad exclusively, so far as it may be practicable, in the transportation of freight and passengers under their control."

By the same token, the Commercial Club of Fairbanks supported the Territorial Highway Commission by further resolving that it endorsed their policy "in arranging its road building program with a view to providing, and maintaining, feeders (roads) for the said branch line." This view is one that is seen today in the current relationship between railroads and the trucking business.

In an attempt to truly co-operate with these concerns, Ohlson ran an eleven day check on the tonnage moving on the Chatanika branch contrasted to that moving by truck during the period from June 25 to July 5. The railroad hauled a little more that 17 tons, while the trucking companies carried 246 tons of freight. Those shipping by truck were some of the leading business establishments in the region, such as the Fairbanks Exploration Company, the Northern Commercial Company and the Standard Oil Company.

Ironically, one of the last bits of freight to be hauled out to the creeks was a steam shovel and the coal needed to fire the engine. It was destined to keep the very roads responsible for the demise of the Chatanika branch in good working order. This unit was a gift from the Alaska Railroad and was brought up from Anchorage to Happy via the standard gauge line, to be transferred at this junction to the narrow gauge line for ultimate shipment out to a new road-camp. This new camp was located at Dome, which served as a road depot for the equipment needed to maintain the roads leading from the Steese Highway.

Fairbanks Daily News-Miner, May 13, 1930 and July 30, 1930

On Friday, August 1, 1930, at 8:00 A.M., Alaska Railroad No. 13-24, a mixed train pulled by engine No. 152, a 4-6-0 Ten Wheeler, left Fairbanks depot for the creeks. Later that day, it returned at 3:30 P.M. from Chatanika as train number 23-14. After leaving the consist in the yard, the locomotive went to the roundhouse, banked its fires for the last time — and, for all intents and purposes brought this skookum railroad to its end!

"Skookum"—An Alaska/Yukon expression for someone, or something, that is good or great.

The Old Reliable Richardson Highway "Takes You Out"

Railroads may come and go, but the Highway goes on, forever. As train service is not asssured before June 25th. We were compelled to get into action on the Highway earlicr than we intended, and our first stage left this morning taking our passengers to catch the "Alaska." The advises from the Engineering Commission are that the mails will be routed over the Richardson Highway, also, until the railroad is repaired. Our stages leave as needed. Make your reservations at E.H. MACK CO., and we will do the rest.

RICHARDSON HIGHWAY TRANSPORTAT

Robert E. Sheldon, Local Manager

Fairbanks Daily News-Miner, June 12, 1923

With the completion of the last run on August 1, 1930, the task of dismantling the line was to be determined at a later date. It was the intention of O.H. Ohlson, General Manager of the Alaska Railroad, not to take up the narrow gauge track for a period of one year in order to re-evaluate the effects of of the loss of the railroad on the community. Further, the tracks were to be kept in place for this period to guard against maintenance failure on the surrounding roads and to protect the Interior against exorbitant trucking charges.

Fairbanks Daily News-Miner, May 13, 1930

Satisfied with the feasibility and success of highways, Ohlson, in 1931, gave the order to have the 30, 35 and 40 pound rails removed from the Chatanika branch. However, acting on a petition from the miners prospecting on the Goldstream, Ohlson granted them their request to leave the bridges and culverts below milepost 18 (Fox) in place when the rails from this line were removed. It was their intent to use the roadbed as a motor highway from Fox down through the Goldstream Valley. As a result, to this very day one can still see two of the remaining bridges in this area, along with a multitude of wooden culverts.

Paragraph 2 condensed from a December 7, 1930 letter written by O.H. Ohlson, General Manager, Alaska Railroad to the Secretary of the Interior, and consolidated with a Fairbanks Daily News-Miner article published July 5, 1932

Line-side railroad structures, such as Gilmore Station, Olnes Station, Little Eldorado Station and other related railroad structures dispersed along the line, were to be subjected to sealed-bid sales. These bids were to be publicly opened at 2:00 P.M. on August 31, 1931, and sold off.

Fairbanks Daily News-Miner, August 8, 1931

By the time the Alaskan Engineering Commission took over the Tanana Valley Railroad, most of the equipment was pretty much worn out. Generally the cars were in a poor and dilapidated state and were of little use to the new management. Some of these narrow gauge cars, dating back from the period when the Alaskan Engineering Commission was building the line, were converted for use on the standard gauge track of the Alaska Railroad. This was accomplished simply by removing the narrow gauge trucks and replacing them with standard gauge ones. Included in one of these transformations was one of the original Tanana Valley passenger cars. All found new service on the Alaska Railroad. Some of the other cars, including an original four wheeler boxcar, were transferred to the Alaska Railroad's Moose Creek operation near Palmer, which was also, originally, a narrow gauge operation. Disposition of the remaining narrow gauge equipment is unknown.

Author's Note:
There has been some feeling among the local historians that the remains of the Tanana Valley Railroad were sold off to Japan prior to World War II. I have been unable to confirm this.

Removal of track was a gradual process. It began with the removal of the unused sidings, the remains of the Chena line and finally, the main line. The existing rail was then recycled for use in the construction of the Alaska Railroad. Since the ties were not creosoted, the local residents obliged themselves by scavenging these pieces and converting them into a much needed source of fuel for their homes and mining operations.

Paragraph 2 on page 143 referernces *Rails North*, pg. 125

However, to this very day one can still find rotting ties in various stages of decomposition, along many of the isolated stretches of the old Tanana Valley Railroad.

Locomotives No. 50, 51 and 52, were sold to the Alaskan Engineering Commission when it took over the Tanana Valley Railroad in 1917. They were used primarily for maintenance work and for the construction of the narrow gauge line from Happy south to North Nenana. They were reportedly scrapped in 1930. Locomotive No. 151, a 2-8-0 Consolidation, was purchased by the Alaskan Engineering Commission, but was used sparingly. It was far too heavy for the marginal conditions of the Tanana Valley Railroad roadbed. Retired in 1921, it was supposedly scrapped in 1936. Ten Wheeler 4-6-0, No. 152, purchased from the Baldwin Locomotive Works, was the only *new* locomotive to have ever worked the lines of the Tanana Valley Railroad. It became the workhorse of the Chatanika branch during its tenure on the Alaska Railroad. It still is in operation today on a Michigan tourist railroad.

And — after standing in the heat, cold, rain and snow for 66 years, first along side the Alaska Railroad's depot in Fairbanks and then subsequently in the town's historical theme park, *Alaskaland*, locomotive No. 1, the first engine on the Tanana Mines Railway, like Lazarus, has risen from the dead! Under the outstanding direction of *Dan Gullickson* of Fairbanks, Alaska, and the *Friends Of The Tanana Valley Railroad*, the locomotive is being rebuilt to an operational status. Come ride with us, for yesterday's train will soon leave the depot, *TODAY*!

Author's Note:

Friends of the Tanana Valley Railroad, Inc., a 501 (c) (3) public nonprofit organization, was incorporated in 1992. The FTVRR was created to restore and operate Tanana Valley Railroad engine #1, (Porter locomotive Works #1972), and construct a living museum dedicated to the Tanana Valley Railroad, which for a time was the farthest north railroad in North America.

FTVRR's international membership is open to anyone and consists of individuals, organizations and businesses sharing a common interest in preserving history. For information, contact Friends of the Tanana Valley Railroad, P.O. Box 84498, Fairbanks, Alaska, 99708; FAX (907) 479-3276; or E-mail ftvrrfun@polarnet.fnsb.ak

A portion of the purchase price of this book is being donated to Friends of the Tanana Valley Railroad, Inc.

Friends of The Tanana Valley Railroad working on locomotive No. 1.
Photo #1 features Curt Fortenberry, and the N.C.Co. forklift. #2 features Jerry Zellmer. The forklift in this picture was loaned by MarkAir. Photo #3 features Ben Anderson (standing on bumper), an unseen member beside him, and Curt Fortenberry. The forklift in use at that time was loaned by the N.C. Co. Photo back cover - Author Nicholas Deely working at dismantling locomotive No. 1 for renovation.
All photos courtesy of the *Friends of The Tanana Valley Railroad*

Unless otherwise indicated, all of the following data was obtained from the Archives of the Alaska Railroad.

APPENDIX A

LOCOMOTIVES, ROLLING STOCK AND STRUCTURES.

The Tanana Mines Railway/Tanana Valley Railroad had four locomotives, 38 freight cars, one electric car and one snowplow that was subsequently converted into a "dozer." Being an isolated railroad in the wilderness of Alaska, with no other physical connection to any other railroad, it had no interline restraints placed on it. As a result, it built and modified the rolling stock to meet its own immediate needs without concern for established standards.

Couplers on both the locomotives and the rolling stock were of the *link* and *pin* variety. This method of coupling cars together was outlawed just as soon as the Alaskan Engineering Commission took over the railroad in 1917. Not only was it a primitive system, but it was responsible for maiming and crushing many a hand. Prior to this take over, the Tanana Valley Railroad could not equip its cars with automatic couplers, as mandated by the Interstate Commerce Commission, for the deplorable state of the *roadbed* could not handle the banging of one car up against the other during the process of coupling the wagons together.

LOCOMOTIVES OF THE TANANA VALLEY RAILROAD.

No. 1:
Type, 0-4-0, tank engine. Built by Porter in March, 1899. Weight - 7 tons. Wheel base - 48". Drivers - 24". Cylinders, 6" x 10". Cab was made out of wood. This was the first locomotive on the Tanana Mines Railway. It still exists today and is being rebuilt into an operational status. *(Mining Railways of the Klondike, Eric L. Johnson, pg. 66)*

No. 50:
Type, 4-4-0 American. Made by Baldwin. Purchased from the White Pass Railway by the Tanana Valley Railroad in 1905. Locomotive weight, 18 tons. Tender weight, 6 tons. Total weight, 24 tons. Cylinders, 11" x 16". Drive wheels, 40" diameter. Tractive effort, 5,400 lbs. Stephenson link valve gear. Total engine wheel base, 21 ft. 7½ inches. Eight wheel tender with steel underframe. Link and pin coupler. Working pressure 125 lbs. Tubes, 113, 1¼ inch, 8 ft. 6 inches long.

No. 51:
Type, 2-6-0 Mogul. Made by Brooks. Purchased from the White Pass Railway by the Tanana Valley Railroad in 1905. Locomotive weight, 25 tons. Tender weight, 7 tons. Total weight, 32 tons. Cylinders, 14" x 18". Drive wheels, 38" diameter. Tractive effort, 9,500 lbs. Stephenson link valve gear. Total engine wheel base, 25 ft. 9 inches. Link and pin coupler. Eight wheel tender with steel frame. Working pressure, 125 lbs. Tubes, 86, 1¼ inch, eight feet long.

No. 52:
Type, 2-6-0 Mogul. Made by Baldwin. Purchased from the Denver and Rio Grande Railroad by the Tanana Valley Railroad in 1906. Locomotive weight, 28 tons. Tender weight, 8 tons. Total weight, 36 tons. Cylinders, 11" x 18". Drive wheels, 38" diameter. Tractive effort, 9,500 lbs. Stephenson link valve gear. Total engine wheel base, 25 ft., 9 inches. Link and pin coupler. Eight wheel tender with steel frame. Working pressure, 125 lbs. Tubes, 86, 1 ¼ inch, 8 ft. long.

LOCOMOTIVES PURCHASED FOR USE ON THE TANANA VALLEY RAILROAD AFTER IT CAME UNDER THE CONTROL OF THE ALASKAN ENGINEERING COMMISSION.

No. 151:
Type, 2-8-0 Consolidation. Purchased by the Alaskan Engineering Commission in 1917 from a logging railroad in Seattle, Washington. Cylinders - 16" x 21". Drive wheels - 36" diameter. Total wheel base - 19 ft. 2½ inches. Link and pin coupler - later changed to an automatic coupler. Working pressure, 185 lbs. Tubes, 144, 2 inch, 10 ft. 8 inches in length. First placed in service, December 9, 1919.

This engine was the largest locomotive on the Tanana Valley Railroad. Although it was used primarily on the North Nenana run, and at times, paradoxically on the *winter ice crossings* on the Tanana River, it was deemed to be unsafe, because of poor track, to operate out on the line to Chatanika. Along with locomotive No. 152, it was also used out along the mainline to the creeks in the winter *when the ground and track were firmly frozen together.* During the summer, both locomotives were taken off the Chatanika line and replaced by the much lighter, but less powerful, locomotive No. 52. Since the grades on this line were very steep, this engine had great difficulty in getting over the hills. *(Letter from J.T. Cunningham, Supt. of Transportation, to Mr. N. Horm, April 22, 1924)*

When locomotive No. 151 was placed in service it was found to have the proper grates for burning coal, a situation not found on the other engines. Not only was this change economical from a fuel standpoint, but two hours or more were saved in loading-up the locomotive tender with wood – piece by piece. Further with coal, there was no need to refuel again anywhere along the line. *(Fairbanks Daily News-Miner, Friday, April 16, 1920)*

No. 152:
Type, 4-6-0 Ten Wheeler. Purchased new from the Baldwin Locomotive Co. in 1920. It was shipped to Nenana by riverboat, in pieces, where it was assembled for use on the Chatanika line. Cylinders 14" x 20". Driver wheel diameter was 44". Weight on drivers 50,000 lbs. It became the "“work-horse” of the Alaska Railroad narrow gauge lines (Tanana Valley Railroad) going on alternate days from Fairbanks to Chatanika and back, and then from Fairbanks to North Nenana and back. Trains were usually of a mixed variety, hauling both passenger and freight cars. It survives today on an amusement park railroad in Michigan. Oddly enough, the only surviving remnants of the Tanana Valley Railroad locomotive power should be its very first unit, No. 1, and its very last bit of motive power, No. 152.

During the construction years of the line from North Nenana to Fairbanks, between 1918 and 1923, the A.E.C. had an additional four other “dinky” narrow gauge locomotives. They were used primarily for the construction of this line.

OTHER LOCOMOTIVES OF THE ALASKAN ENGINEERING COMMISSION.

Engines No. 4 & 6:
Type, 0-4-0, four wheel switcher, similar to Tanana Valley's famous No. 1 locomotive. Cylinders 7" x 12". Wheel diameter 23". Drive wheel base 4 ft. 9 in. weight 30,000 lbs. Stephenson link valve gear. Four wheel tender — for coal only. The sides and ends were built of wood. Saddle tank for water. Link and pin coupler. Working pressure 110 lbs. 48, 1½ inch tubes, 8 ft. 4 in. long. By 1924 they were both partly stripped for parts and the boilers used for a water tank and reservoir, respectively.

No. 21 & 22:
Built by the Davenport Locomotive Works in 1908 at a cost of $3,163 each. Type 0-4-0. Cylinders - 10" x 16". Driving wheels diameter - 29 inches. Tractive effort - 5,620 lbs. Hauling capacity on a level surface - 730 tons. Stephenson link valve gear. Driving wheel base - 4 feet, 2 inches. Saddle water tank with a capacity of 607 gallons. Coal - 1000 lbs. Link and pin coupler. Boiler pressure - 130 lbs. These two locomotives were leased to coal contractors whose operations were located along the 3 foot Moose Creek narrow gauge branch of the Alaska Railroad. This branch was an extension of the standard gauge Palmer branch.

No. 50: *This was the second locomotive numbered 50 to be used on this line.*
Type, 0-6-0. Built by H.K. Porter Company in 1910. Cost, $2,905.00. Cylinders, 8" x 14". Diameter of driving wheels, 24". Weight 23,000 lbs. Tractive effort 4,400 lbs. Hauling capacity, 630 tons. Stephenson link valve gear. Driving wheel base, 5 ft. 6 inches. Water capacity, 350 gallons. Coal, 400 lbs. Working pressure 160 lbs. Tubes, 67, 1½ inch. Used frequently on the Nenana - North Nenana run in the winter, across the ice when this frozen mass appeared too weak to support a heavier locomotive. Scrapped by 1924.

PASSENGER CARS.

During the years of its operation, the Tanana Valley Railroad had four passenger cars in service. By 1918, when the line was taken over by the Alaskan Engineering Commission, it possessed only three of these cars. As best as can be ascertained, the disposition of these cars were as follows:

No. 200:
Length - 40 feet with a seating capacity of 44. It continued its service to the Interior beyond the years of Tanana Valley Railroad ownership. It was placed in service on the Alaska Railroad and probably "died of old age" like so much of the other equipment.

No. 202:
Length - 49 feet with a seating capacity of 44. It was rebuilt in the Ship Creek (Anchorage) shops into a combination passenger - baggage car. *(Alaska Railroad Annual Report, 1918)*

No. 203:
Length - 49 feet with a seating capacity of 44. The car was repaired and rebuilt, *but with standard gauge trucks*, for use as a gas car trailer. *(Alaska Railroad Annual Report, 1917)*

No. 204:
Originally, this coach came from the White Pass and Yukon Railway. In 1916, just before the Tanana Valley Railroad was taken over by the Alaskan Engineering Commission, it was destroyed in an accident at Merino when the whole train "jumped" the trestle at that location.

FREIGHT CARS AS OF MAY 10, 1918.

Flatcars:
Numbers 51, 53, 59 & 61 were built by the White Pass and Yukon Railway for the Klondike Mines Railway located in the Yukon Territory of Canada. However, because of the fluctuating fortunes of this railway, they were never picked up by the line. Like so much of the equipment on the Tanana Mines Railway / Tanana Valley Railroad that came from all corners of the continent, these four cars were purchased by this company in 1906 and were placed on its own lines without bothering to change the reporting marks on the sides of these cars. Thus, one may see photographs of the Tanana Valley locomotives pulling flatcars lettered for the Klondike Mines Railway.

Numbers 919 & 920 were converted into boxcars and were re-numbered No.'s 303 & 304, respectively. No. 921 was converted into *caboose No. 1052. (Letter from R.B. Ward, Acting Division Engineer, Alaskan Engineering Commission, to Wm. Gerig, Engineer in Charge, May 10, 1918)*

MISCELLANEOUS.

Electric Car with storage batteries — "Federal". Seating capacity 25. Good Condition. One Rheostat for charging electric car - 120 volts. One Volt meter - 120 volts. Asking selling price was $750.

One - 6 cylinder "Buda" motor car - gasoline - 50 H.P. Seating capacity 17 passengers.
One - 4 cylinder "Fairbanks - Morse" gasoline 12 H.P. Motor car, air cooled. Seating capacity 7. "Speeder".
One - 2 cylinder "Fairbanks - Morse" gasoline motor car, 14 H.P. Seating capacity 7.
One - 1 cylinder "Fairbanks - Morse" air cooled motor car. Seating capacity 3 - bad order.

Dozer attached to flatcar #57, 28 feet double truck, snowplow. In need of repairs.

Automobile chassis and body with wheels, no tires. Bad condition.

TANANA VALLEY RAILROAD FREIGHT CAR EQUIPMENT.

Boxcars #60, 61, 62, 63, 64 & 65 All had the same measurements. These were the "shorty" four wheel cars.

Length inside	18 ft. 1 in.	
" outside	19 ft.	
Width inside	6 ft. 5 in.	
" outside	7 ft. 6 in.	(Level full 744.328 / cu. ft., each)

Height inside	6 ft.	2 in.	
" outside from top of rail	9 ft.	6 in.	
Extreme height top of running board	10 ft.		
Height of doors	5 ft.	7 in.	
Width of doors	4 ft.		

Boxcar #114

Length inside	30 ft.	6 in.	
" outside	31 ft.	4 in.	
Width inside	7 ft.		
" outside	7 ft.	10 in.	
Height inside	6 ft.	2 in.	(1316.441 / cu. ft.)
" outside from top of rail	9 ft.	8 in.	
Extreme height top of running board	10 ft.	4 in.	
Height of doors	6 ft.		
Width of doors	5 ft.		

Boxcar #118

Length inside	24 ft.	3 in.	
" outside	25 ft.		
Width inside	7 ft.		
" outside	7 ft.	8 in.	
Height inside	6 ft.	6 in.	(Level full 1193.375 / cu. ft.)
" outside from top of rail	10 ft.	2 in.	
Extreme height top of running board	11 ft.		
Height of doors	6 ft.	2 in.	
Width of doors	5 ft.	6 in.	

Boxcar #106

Length inside	24 ft.	3 in.	
" outside	25 ft.		
Width inside	6 ft.	6 in.	
" outside	7 ft.	8 in.	
Height inside	6 ft.	6 in.	(Level full 1103.375 / cu. ft.)
" outside from top of rail	7 ft.	8 in.	
Extreme height top of running board	11 ft.		

Height of doors	6 ft. 2 in.	
Width of doors	4 ft. 6 in.	

Boxcar #110

Length inside	29 ft. 1 in.	
" outside	30 ft. 2 in.	
Width inside	6 ft. 2 in.	
" outside	7 ft. 7 in.	
Height inside	6 ft. 2 in.	(1231.488 / cu. ft.)
" outside from top of rail	9 ft. 8 in.	
Extreme height top of running board	10 ft. 6 in.	
Height of doors	5 ft. 10 in.	
Width of doors	4 ft. 8 in.	

Flatcars # 117, 119, 121 & 123 Same Measurements.

Length	30 ft.
Width inside	7 ft.
Height from rail	2 ft. 10 in.
Extreme height top of car	4 ft.

Flatcar # 113

Length	31 ft.
Width inside	7 ft. 4 in.
Height from rail	2 ft. 4½ in.
Extreme height top of car	3 ft. 2½ in.

Flatcar # 115

Length	31 ft. 5 in.
Width inside	7 ft. 4 in.
Height from rail	2 ft. 8 in.
Extreme height top of car	3 ft. 6 in.

Flatcar # 103

Length	32 ft. 2 in.
Width inside	7 ft. 4 in.
Height from rail	2 ft. 8 in.

Extreme height top of car	3 ft.	6 in.
Flatcar # 107		
Length	32 ft.	2 in.
Width inside	7 ft.	4 in.
Height from rail	2 ft.	8 in.
Extreme height top of car	3 ft.	6 in.
Flatcar # 109		
Length	31 ft.	4 in.
Width inside	7 ft.	2 in.
Height from rail	2 ft.	6 in.
Extreme height top of car	3 ft.	5 in.
Flatcar # 225		
Length	30 ft.	8 in.
Width inside	7 ft.	6 in.
Height from rail	2 ft.	3 in.
Extreme height top of car	3 ft.	2 in.
Flatcar # 233		
Length	30 ft.	4 in.
Width inside	7 ft.	
Height from rail	2 ft.	1 in.
Extreme height top of car	2 ft.	11½ in.
Flatcar # 111		
Length	31 ft.	1 in.
Width inside	7 ft.	1 in.
Height from rail	2 ft.	6 in.
Extreme height top of car	3 ft.	6 in.
Boxcar #112		
Length inside	26 ft.	2 in.
" outside	27 ft.	
Width inside	7 ft.	

" outside	7 ft.	6 in.	
Height inside	6 ft.	4 in.	(Level full 1159.96 / cu. ft.)
" outside from top of rail	10 ft.	2 in.	
Extreme height top of running board	11 ft.		
Height of doors	6 ft.	2 in.	
Width of doors	4 ft.	6 in.	

Boxcar #108

Length inside	30 ft.	6 in.	
" outside	7 ft.	8 in.	
Width inside	7 ft.		
" outside	7 ft.	8 in.	
Height inside	6 ft.	5 in.	(Level full 1374.816 / cu. ft.)
" outside from top of rail	10 ft.	2 in.	
Extreme height top of running board	10 ft.	6 in.	
Height of doors	6 ft.		
Width of doors	5 ft.		

Boxcar #116

Length inside	26 ft.		
" outside	27 ft.		
Width inside	7 ft.		
" outside	7 ft.	9 in.	
Height inside	6 ft.	4 in.	(Level full 1152.606 / cu. ft.)
" outside from top of rail	10 ft.	2 in.	
Extreme height top of running board	10 ft.	10 in.	
Height of doors	6 ft.	1 in.	
Width of doors	4 ft.	6 in.	

Boxcar #57 (Dozer)

Length inside	25 ft.	9 in.	
" outside	26 ft.	10 in.	
Width inside	7 ft.	1 in.	
" outside	7 ft.	10 in.	
Height inside	6 ft.	4 in.	(Level full 1171.319 / cu. ft.)
" outside from top of rail	10 ft.	2 in.	

Extreme height top of running board	10 ft. 10 in.
Height of doors	6 ft. 3 in.
Width of doors	2 ft. 2 in.

Boxcars were painted a metallic red while the flatcars were a carbon black. When the Tanana Valley Railroad was taken over by the Alaskan Engineering Commission, all boxcars had a bold U.S. painted on their sides with the appropriate car number. Both were 8 inches high. *(Letter from Chairman Edes, Alaskan Engineering Commission, to Fredrick D. Brown, Engineer in charge, Nenana, Alaska, October 1, 1918)*

APPENDIX B

STRUCTURES.

Fairbanks Depot:
30 x 50 ft., double story. Downstairs divided into public waiting room, ticket office and baggage room. Upstairs divided into offices and living rooms. *(August 5, 1905-Fairbanks Weekly News)*

Gilmore Station:
One and one-half stories frame structure, tin roof, shiplap sides lined with beaver boards, seven rooms, size 25' x 45' with lean-to 12' x 12'.

Gilmore Warehouse:
One and one-half stories, 12' x 60', wood frame building with a corrugated iron roof, and a loading platform 12' x 60', 4 feet high.

Wagner Section House:
Log structure,18' x 30', with 9' x 40' outfit car addition. It had three rooms.

Olnes Station:
Three room frame structure, 25' x 40', corrugated iron roof.

Little Eldorado Station:
Frame structure,16' x 75', with iron roof.

Section House at Mile 26, bridge No. 5:
Frame structure, 18' x 52', iron roof and sides lined with shiplap and covered with building paper. Adjacent shed (old outfit car) 9' x 40'. *(Fairbanks News-Miner, August 19, 1932)*

APPENDIX C

BRIDGES, CULVERTS AND LOCATION:

In their haste to construct the Tanana Mines Railway, and then the Tanana Valley Railroad, the engineers took some short cuts in order to get the trains into operation as soon as possible. Between every piling, that is the vertical uprights of a bridge, there were horizontal spans called stringers, which were to be no longer than 14 feet. While this engineering principal was adhered to in many cases, there were exceptions for frequently spans were found to measure 16 feet. Further, instead of having three of these 10" x 12" stringers under each rail, the bridges were built with only two. Although there is no known history of any bridge collapsing under the weight of a train, serious breaches in established engineering basics compromised the longevity of these structures.

Fairbanks to Chena Junction:

Mile - 0.2, pile and frame trestle, over slough	167' 0".
Mile - 0.9, pile and frame trestle, over slough	126' 0".
Mile - 1.1, open culvert	8' 0".
Mile - 1.3, open culvert	8' 0".
Mile - 2.2, pile and frame trestle over Noyes Slough	362' 0".
Mile - 2.6, pile and frame trestle over slough	91' 0".
Mile - 3.2, open culvert	5' 0".
Mile - 3.4, open culvert	12' 0".
Mile - 4.8, open culvert	5' 0".
Mile - 4.9, open culvert	22' 0".
Mile - 5.3, frame trestle, St. Patrick's Creek	45' 0" long.
Mile - 5.6, pile and frame trestle, St. Patrick's Creek	38' 0" long.
Mile - 6.1, open culvert	5' 0" long.
Mile - 6.2, open culvert	10' 0" long.
Mile - 6.3, open culvert	5' 0" long.
Mile - 6.6, open culvert	6' 0" long.

Chena Spur (Chena Junction to Chena):

Mile - 0.1A, open culvert	12' 0" long.
Mile - 0.1A, open culvert	8' 0" long.
Mile - 0.4A, frame trestle over Criple Creek	186' 0" long.
Mile - 2.7A, open culvert	10' 0" long.
Mile - 3.1A, open culvert	10' 0" long.
Mile - 3.2A, open culvert	12' 0" long.

Mile - 3.5A, open culvert	12' 0" long.
Mile - 3.7A, pile trestle, over slough	124' 0" long.
Mile - 3.9A, open culvert	6' 0" long.

Sheep Station (Happy) to Gilmore:

This 13.2 mile section of the line had 67 openings of various lengths. All had to be strengthened to meet the weight limits set by the 32 ton, 4-6-0, Baldwin locomotive. The strain which would be produced by this locomotive on the stringers would be nearly 900 pounds per square inch, which was more than should have been allowed for Alaska spruce. The repairs on the rotted stringers, pilings, cross bracers and sills had to be considered at best, temporary, for the bridges in their entirety eventually had to be placed.

Mile - 9.4, frame trestle	28' 0" long.
Mile - 9.4, frame trestle over Goldstream	74' 0" long.
Mile - 16.8, frame trestle	30' 0" long.
Mile - 16.9, frame trestle	45' 0" long.
Mile - 18.0, frame trestle, Station Fox	58' 0" long.
Mile - 19.2, frame trestle	28' 0" long.
Mile - 19.7, frame trestle	44' 0" long.
Mile - 20.0, frame trestle over Goldstream	44' 0" long.
Mile - 20.3, frame trestle over Goldstream	56' 0" long.

Gilmore to Olnes:

This was the 13.2 mile hill section of the line. It was here that the largest and longest bridges of the Tanana Valley Railroad could be found.

Mile - 20.5	12' long.
Mile - 20.7	16' long.
Mile - 21.1	8' long.
Mile - 21.1	11' long.
Mile - 21.2	19' long.
Mile - 21.2	10' long.
Mile - 21.3	6' long.
Mile - 21.3	6' long.
Mile - 21.4, frame trestle	403.' long.

Ties on the bridge were 6" x 8" x 9'. Every sixth tie was spiked down onto the bridge. Guard rails were bolted every 12' or less. Ties were laid down on 2' centers.

Mile - 21.5	10' long.
Mile - 22	12' long.
Mile - 22.1	6' long.
Mile - 22.1	6' long.

Mile - 22.4	10' long.
Mile - 22.4	6' long.
Mile - 22.4	14' long.
Mile - 22.5	10' long.
Mile - 22.6	12' long.
Mile - 22.6	6' long.
Mile - 22.7	11' long.
Mile - 22.9	13' long.
Mile - 23.0	6' long.
Mile - 23, frame trestle	264' long.

These large trestles had little platforms which extended out from the side of the structure at track level. Barrels full of water were perched on these platforms so as to offer a ready supply of a fire repellent should sparks from a passing locomotive ignite any of the timbers. By the time that the Alaskan Engineering Commission took over the Tanana Valley Railroad these barrels were either broken or absent.

Mile - 23.1	14' long.
Mile - 23.2	12' long.
Mile - 23.3	10' long.
Mile - 23.4	12' long.
Mile - 23.5	6' long.
Mile - 23.6	11' long.
Mile - 23.8	15' long.
Mile - 23.9	10' long.
Mile - 24.0	13' long.
Mile - 24.1, frame trestle	222' long.
Mile - 24.2	11' long.
Mile - 24.3	13' long.
Mile - 24.4	12' long.
Mile - 24.6, frame trestle	428' long.

The deck in this trestle, as with all previously mentioned trestles, was built too light in relationship to the established standards for bridge construction. Braces were *spiked on*, where as, they should have been *bolted on*.

Mile - 24.7	12' long.
Mile - 24.9	12' long.
Mile - 25.1, frame trestle	502' long.

Forty feet high at the deepest part. Some of the ties were only 6" x 6" x 9' instead of the standard of 6" x 8" x 9'. Deck was built too lightly. Very few of the ties were fastened to the stringers. The guard rails, 6" x 6", were notched over the ties, and bolted down only on every 6th tie.

Mile - 25.2	6' long.

Mile - 25.3	6' long.
Mile - 25.4	6' long.
Mile - 25.4	6' long.
Mile - 25.5	6' long.
Mile - 25.6	6' long.
Mile - 25.6	10' long.
Mile - 25.7	26' long.
Mile - 25.8	8' long.
Mile - 26.4	13' long.
Mile - 26.5	6' long.
Mile - 26.6	13' long.
Mile - 26.6	14' long.
Mile - 26.7	6' long.
Mile - 26.7	6' long.
Mile - 26.8	6' long.
Mile - 26.8	14' long.
Mile - 26.9	12' long.
Mile - 27.2	13' long.
Mile - 27.7	7' long.
Mile - 27.8	6' long.
Mile - 27.9	13' long.
Mile - 28.0	13' long.
Mile - 28.2	13' long.
Mile - 28.4	12' long.
Mile - 28.5	6' long.
Mile - 28.9	9' long.
Mile - 29.0	12' long.
Mile - 29.1	6' long.
Mile - 29.2	14' long.
Mile - 29.4	13' long.
Mile - 29.5	12' long.
Mile - 29.6	12' long.
Mile - 29.7	8' long.
Mile - 29.9	12' long.
Mile - 30.0	12' long.
Mile - 30.1, frame trestle	220' long.

Ties were loose. No guard rails were present — replaced by narrow strips of round saplings nailed to the ties.

Mile - 30.3, frame trestle 431' long.

Ties were in a poor condition on this trestle (rotting). No guard rails were present — replaced by small sapplings which were nailed to the ties. Only 4 water barrels were in place.

Mile - 30.4, frame trestle 168' long.
Mile - 30.6 6' long.
Mile - 30.6 8' long.
Mile - 30.8 13' long.
Mile - 31.0 6' long.
Mile - 31.1 8' long.
Mile - 31.2 6' long.
Mile - 31.3 9' long.
Mile - 31.4 6' long.
Mile - 31.4 6' long.
Mile - 31.5 6' long.
Mile - 31.6 6' long.
Mile - 31.7 6' long.
Mile - 31.8 6' long.
Mile - 31.8, frame trestle 335' long.
Mile - 31.9 14' long.
Mile - 32.0 14' long.
Mile - 32.1, trestle 41' long.
Mile - 32.1 6' long.
Mile - 32.4 6' long.
Mile - 32.5 6' long.
Mile - 32.5 6' long.
Mile - 32.8 13' long.
Mile - 32.8, two spans, 8' each 17' long.
Mile - 32.9 17' long.
Mile - 32.9 17' long.
Mile - 33.0, frame trestle 319' long.

Over Dome Creek. One of the few bridges on the T.V.R.R., in 1917, in relatively good shape.

Mile - 33.1 14' long.
Mile - 33.2 6' long.
Mile - 33.3 13' long.
Mile - 33.4 6' long.

Mile - 33.4 6' long.
Mile - 33.5 6' long.
Mile - 33.6, frame trestle 33 ' long.

The engineers of the Alaskan Engineering Commission could not determine if this bridge had two or three spans. If it was a two span bridge, the spans were too long to safely support the weight of the train although it had been doing this for the last 12 years.

Mile - 33.7 14' long.

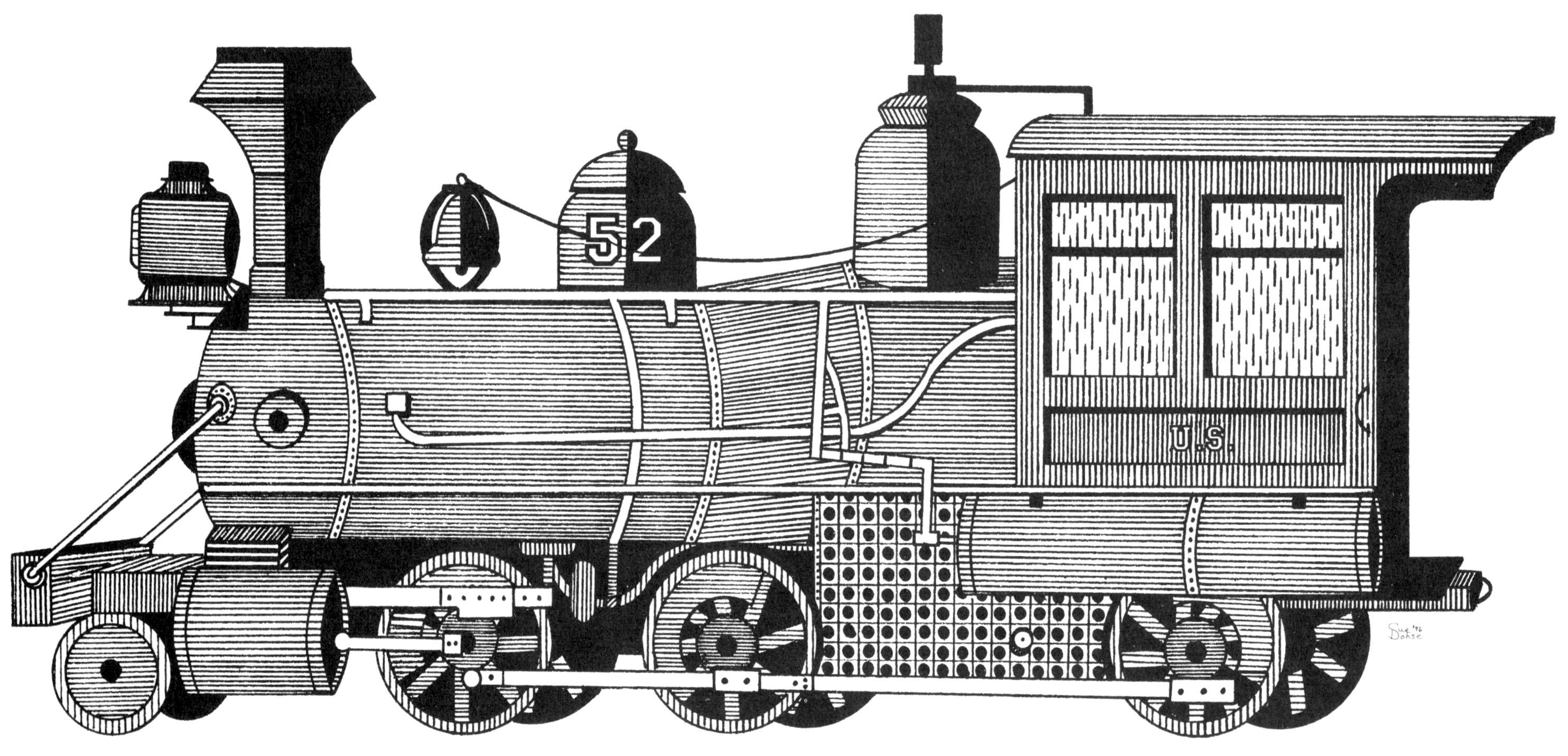

APPENDIX D

Olnes to Chatanika

This section of the line was 5.5 miles long. It ran through swampy terrain.

Mile - 35.7, frame trestle over Little Eldorado Creek 24' long.

Mile - 38.1, frame trestle over Ruby Creek 68' long.

There were 36 other openings on this section of the line which were under 6 feet in length. Gradually they were being inundated by the surrounding swamp land.

The Alaskan Engineering Commission found, upon taking over the Tanana Valley Railroad, that many of the bridges were improperly constructed and were beginning to rot away. The following recommendations were made and carried out:

1. Decks of all the bridges must be strengthened.
2. Mudsills and bearing sills, on the smaller bridges, had to be fixed, knowing fully well, however, that this would give them only temporary relief.
3. The 10" stringers between the vertical bents were too small and insufficient in number. These stringers would have to be replaced and increased in number.

Since the decay and poor construction practices of the bridges were so extensive, it was elected by the Alaskan Engineering Commission to reconstruct the line, up through the hills, by developing a series of "cuts and fills". By doing this, there would be a greater stability to the roadbed. The weight of a passing train was more equally distributed over a greater surface. Therefore, a wider shelf, or "cut" was made into the side of the hills. The diggings from these "cuts" were then used to fill in the larger trestles, thus eliminating the need to repair them. Only the bridge at Dome, two miles south of Olnes, was retained due to geological and terrain difficulties. This bridge was entirely replaced by a new and more substantial one. *(Letter from W. Fogelstrom, Bridge Engineer, Alaskan Engineering Commission, to Thomas Riggs, Jr., Commissioner, Alaskan Engineering Commission, Nenana, Alaska, September 19, 1917)*

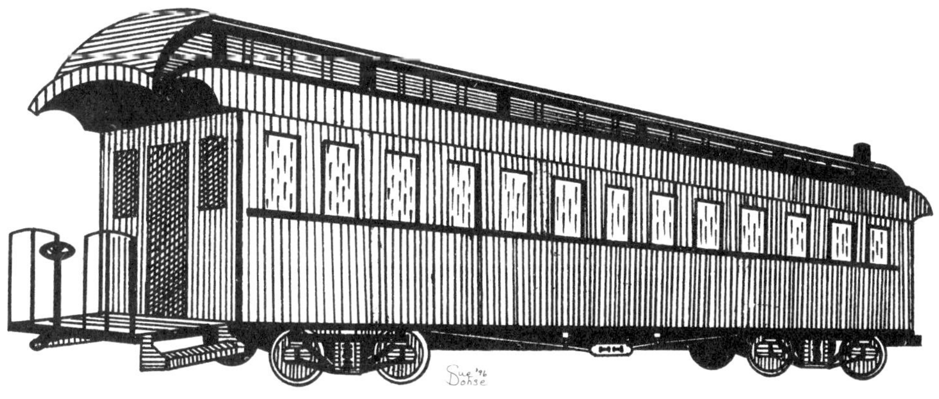

A

B

Italicized numbers indicate pages with photographs / **Bold** numbers indicate entries with comprehensive references.

G

H

I

J

K

L

M

N

O

P

Italicized numbers indicate pages with photographs / **Bold** numbers indicate entries with comprehensive references.

U

V

W

Italicized numbers indicate pages with photographs / **Bold** numbers indicate entries with comprehensive references.

White, G. - Zellmar, J.

X, Y & Z